Recent Advances Towards Improved Phytoremediation of Heavy Metal Pollution

Edited By

David W.M. Leung

School of Biological Sciences
University of Canterbury
Private Bag 4800
Christchurch 8140
New Zealand

CONTENTS

FOREWORD

The idea of decontamination of environments using green plants is not new. Almost 300 years ago, plants were proposed to be used in the treatment of wastewater. At the end of the 19th century, *Thlaspi caerulescens* and *Viola calaminaria* were the first plant species documented to accumulate high levels of metals in leaves. At present, there are about 420 species belonging to about 45 plant families which have been reported as hyperaccumulators of heavy metals.

Nowadays, we know much more about the mechanism of plant tolerance towards heavy metal removal and plant efficiency in uptake, translocation, and further sequestration of heavy metals in specialized tissues (in trichomes and organelles such as vacuoles). It is clear enough that uptake of metals depends on their bioavailability, and also plants have evolved mechanisms to make micronutrients bioavailable; in some cases plants have developed resistance to high metal concentrations, basically by two mechanisms, avoidance and tolerance. In avoidance mechanism, exclusion of metals outside the roots is achieved, and in tolerance mechanism the main undertaking consists basically of complexing the metals to prevent protein and enzyme inactivation. One of the major consequences of heavy metal (HM) accumulation in the cell could be the enhanced generation of reactive oxygen species (ROS) which usually damage the cellular components such as membranes, nucleic acids, chloroplast pigments and alteration in enzymatic and non-enzymatic antioxidants. Interestingly, a new family of molecules designated a reactive nitrogen species (RNS) starts to be new elements involved in the mechanism of response against HM where molecules such as nitric oxide (NO), peroxynitrite (ONOO-) and *S*-nitrosoglutathione (GSNO) can mediate protein function by specific post-translational modifications.

Given the nature and extent of contamination worldwide and the costs involved in remediation, scientists made a drive towards alternative cost effective technologies for the remediation of polluted sites. In this regard, bioremediation, typically referring to microbe-based clean-up, and phytoremediation, or plant-based clean-up, have generated much interest as effective low-cost and environmentally-friendly technologies for the clean-up of a broad spectrum of hazardous organic and inorganic pollutants.

Plant-based environmental remediation has been widely pursued by scientists as a favorable low-cost clean-up technology applicable in both developed and developing nations in recent years. Physiological, biochemical, and molecular approaches are continually being applied to identify the underlying mechanisms of metal tolerance and hyperaccumulation. The drive to find genes underlying these unique biological properties is partly fueled by interest in using transgenic plants in phytoremediation. Interestingly, as transgenics are being tested in the field and the associated risks assessed, their use appears to be more accepted and less regulated than has been the case for transgenic crops.

In last two and half decades, phytoremediation work got so much attention from the scientists and researchers throughout the globe. It is a really promising method to clean our polluted environment. I personally owe debt of gratitude to Dr. David W.M. Leung, who did an excellent job in compiling this eBook with recent advancement in this field and also thankful to the authors who contributed their time and knowledge in making this eBook in a nice shape. Overall the information compiled in this eBook will bring together uesful knowledge and advancement of phytoremediation technologies in recent years.

Dharmendra Kumar Gupta
Departamento de Bioquímica
Biología Celular y Molecular de Plantas
Estación Experimental del Zaidín (EEZ)
Consejo Superior de Investigaciones Científicas
E-18008 Granada
Spain

PREFACE

A global environmental and public health issue is heavy or toxic metal pollution. Use of green plants to clean up heavy metal pollution is an environmentally friendly as well as a low-cost approach to the problem. This plant-based biotechnology to help to better manage this global public health concern is commonly known as phytoremediation. Presently, there is no wide-spread application of this technology because useful plants with enhanced resistance/tolerance to metal toxicity are still needed to assist remediation of toxic metal-contaminated environments. A key to improved phytoremediation of heavy metal pollution lies in research seeking for a better understanding of the mechanism(s) of heavy metal resistance/tolerance in plants.

This eBook is divided into two parts. In the first part, an introduction and an overview of the broad applications of phytoremediation were provided (Chapter 1), particularly useful for senior undergraduate, postgraduate students and non-plant discipline-based researchers interested in environmental biotechnology. This was followed by a closer examination of those research foci evaluating chemicals directly or indirectly (those released by microorganisms) for promoting phytoremediation potential of plants that are not commonly considered as model experimental plants systems (Chapters 1 to 3). In the second part, recent insights, largely at the molecular and genetic level, gained from the use of each of the several model experimental plants in relation to the general theme of improving phytoremediation potential of plants were reviewed in Chapters 4 to 7. The unique arrangement and treatment of the topics have never been brought together comprehensively in a single advanced resource book.

The different chapters were written by researchers who have recently contributed original research papers in the field of phytoremediation. Their contributed chapters in this eBook are, however, written at the levels intended to be useful to students (senior undergraduate and postgraduate), and researchers in plant physiology and biotechnology. In addition, soil scientists, environmental science students and researchers in environmental and contemporary natural resource engineering departments should also find this as a helpful resource.

David W.M. Leung
School of Biological Sciences
University of Canterbury
Private Bag 4800
Christchurch 8140
New Zealand

List of Contributors

Asha Juwarkar, Eco-Restoration Division, National Environmental Engineering Research Institute, Nehru Marg, Nagpur - 440020, India

Bijaya K. Sarangi, Environmental Biotechnology Division, CSIR-National Environmental Engineering Research Institute, Nehru Marg, Nagpur – 440020, India

Brett Robinson, Department of Soil Science, Lincoln University, PO Box 7647, Lincoln, New Zealand

David W.M. Leung, School of Biological Sciences, University of Canterbury, Private Bag 4800, Christchurch 8140, New Zealand

Ian McIvor, Plant and Food Research, Private Bag 11030, Palmerston North, New Zealand

Katarina Vogel-Mikuš, Biotechnical faculty, Dept. of Biology, Večna pot 111, SI-1000 Ljubljana, Solvenia

Lingjuan Zheng, Department of Organismic Biology, University of Salzburg, Hellbrunnerstraße 34, 5020 Salzburg, Austria

Pulavarty Anusha, Environmental Biotechnology Division, CSIR-National Environmental Engineering Research Institute, Nehru Marg, Nagpur – 440020, India

Radha Rani, Eco-Restoration Division, National Environmental Engineering Research Institute, Nehru Marg, Nagpur - 440020, India

Ram Awatar Pandey, Environmental Biotechnology Division, CSIR-National Environmental Engineering Research Institute, Nehru Marg, Nagpur – 440020, India

Sarita Tiwari, Environmental Biotechnology Division, CSIR-National Environmental Engineering Research Institute, Nehru Marg, Nagpur – 440020, India

Thomas Peer, Department of Organismic Biology, University of Salzburg, Hellbrunnerstraße 34, 5020 Salzburg, Austria

Ursula Lütz-Meindl, Department of Cell Biology, University of Salzburg, Hellbrunnerstraße 34, 5020 Salzburg, Austria

DEDICATION

This eBook project was initiated, brutally suspended, restarted progressively and finally completed in Christchurch located within 10 to 20 kilometers from the epicenters of more than 10,000 earthquakes and aftershocks including the killer one in which over 180 people tragically lost their lives in 2011. Much of the Christchurch city needs to be re-built or remediated. Through these all, my wife Irene and my son Joshua, gave me the strength to work on this eBook project among other personal and professional commitments. I would gratefully dedicate this completed eBook to them to mark this unique period of our lives.

CHAPTER 1

Interactions Between Plant Growth Promoting Microbes and Plants: Implications for Microbe-Assisted Phytoremediation of Metal-Contaminated Soil

Radha Rani[*] **and Asha Juwarkar**

Eco-Restoration Division, National Environmental Engineering Research Institute, Nehru Marg, Nagpur-440020, India

Abstract: This chapter first gave a broad overview of the application of phytoremediation technologies for the management of metal-contaminated sites. Then the interactions between plants and the microorganisms in the rhizosphere are reviewed as these could influence the potential accomplishment of these phytotechnologies. Plant-microbe interactions can be enhanced or modulated by modifying microbial population (rhizoengineering) for the remediation of pollutants present in the soils. Rhizoengineering is an innovative approach towards phytoremediation.

Keywords: Bioaccumulation, bioavailability (of heavy metals), bioremediation, biosorption, ecotoxicity, indole-3-acetic acid, metal uptake, mycorrhizal associations, phytoextraction, phytoremediation, phytosiderophore, phytotechnologies, plant growth promoting bacteria, pollution, rhizobacteria (interactions between plants), rhizoengineering, rhizofiltration, rhizosphere, root exudates, toxic metals.

GENERAL INTRODUCTION

With intense industrial and agricultural activities worldwide, contamination of soil with heavy metals has been on a continuous rise, leading to significant health problems and toxic effects on plant and microbial biodiversity. Generally, metals are not degraded biologically or chemically but persist in the environment indefinitely. Consequently, once accumulated these toxic metals render the soil unsuitable for vegetation. Remediation of metal-contaminated soils thus becomes

*Address correspondence to **Radha Rani:** Eco-Restoration Division, National Environmental Engineering Research Institute,Nehru Marg, Nagpur-440020, India; E-mail: raadharaani1982@gmail.com

important. Traditional methods for heavy metal decontamination include excavation, landfill dumping, thermal treatment, acid leaching, and electro-reclamation. However, because of the high cost, low efficiency, and large destruction of soil structure and fertility, these methods are either ineffective or not ecologically sustainable. In order to eliminate or control hazardous chemicals, biological processes are being investigated as alternative approaches. Recently, phytoremediation has emerged as a cost-effective, environment-friendly cleanup alternative that employs the use of higher plants for the cleanup of contaminated environments. Selected plant species possess the genetic potential to remove, degrade, metabolize, or immobilize a wide range of contaminants. The success of phytoremediation depends on the extent of soil contamination, bioavailability of the metal, and the ability of the plant to absorb and accumulate metals in shoots. Plants with exceptionally high metal accumulating capacity often have a slow growth rate and produce limited amounts of biomass when the concentration of metal in the contaminated soil is very high and toxic. To maximize the chance of success for phytoremediation, beneficial microorganisms like plant growth promoting rhizobacteria (PGPR) and arbuscular mycorrhizal fungi (AMF) that inhabit the rhizosphere, are utilized in the nutrient poor soils. They increase heavy metal sequestration capacity of plants by recycling nutrients, maintaining soil structure, detoxifying chemicals, and controlling pests while decreasing toxicity of metals by changing their bioavailability. In return, plants provide the microorganisms with root exudates such as free amino acids, proteins, carbohydrates, alcohols, vitamins or hormones, which are important sources of nutrients for these microorganisms in the rhizosphere. The microorganisms in the rhizopshere interact with each other and with plants and these interactions can greatly influence the success of phytoremediation. To sum up, it is beneficial to exploit the competence of plants and microbes to adapt in the metal-polluted environment and to detoxify toxic metals symbiotically for the successful application of phytoremediation. Therefore, this chapter not only reviews the application of phytoremediation technologies for the management of metal-contaminated sites but also on the interactions between plants and the microorganisms in the rhizosphere as these could influence the potential accomplishment of these phytotechnologies.

SOURCES OF TOXIC METALS IN SOIL AND WATER

'Heavy metal' though used widely in the non-technical and scienetific literature is not an appropriate term as it includes transition metals, metalloids, lanthanides and actinides. However, elements included in the 'heavy metal' category are generally of high atomic number and pose toxic effects to the biota. Thus, an alternative term 'toxic metal' for which no consensus of its exact definition exists, may also be employed. Toxic metals that have been identified in the polluted environment include As, Cu, Cd, Pb, Cr, Ni, Hg and Zn.

Contamination of terrestrial and aquatic ecosystems by toxic metals is a worldwide issue. All countries have been affected, though the area and severity of pollution vary enormously. In Western Europe, over 300,000 fields were contaminated, and the estimated total number in Europe could be much larger, as pollution problems has increasingly occurred in Central and Eastern European countries [1]. In USA, there are 600,000 brown fields which are contaminated with metals and need reclamation [2]. More than 100,000 ha of cropland, 55 000 ha of pasture and 50,000 ha of forest have been lost. The problem of land pollution is also a great challenge in China, where one-sixth of total arable land has been polluted by toxic metals, and more than 40% has been degraded to varying degree due to erosion and desertification. Soil and water pollution is also severe in India, Pakistan and Bangladesh, where small industrial units are pouring their untreated effluents in the surface drains spreading over near agricultural fields. High levels of toxic metals were detected in Yamuna river sediments from Delhi and Agra urban cities, in a study [3].

Sources of toxic metals in soil include both natural and anthropogenic activities. Many of these metals are present in earth's crust naturally, and long range pollution is caused due to volcanic eruptions, forest fire and dust storms. Anthropogenic activities are associated with industrialization and agriculture like waste disposal, atmospheric deposition, waste incineration, industrial effluent, vehicle exhaust and fertilizer and pesticide application. Unlike organic pollutants, metals are not subjected to degradation and hence remain in the environment for a long period of time. During the course of time they have a tendency to bioaccumulate and biomagnify and get entry to the food chain and hence whole biota and cause hazards to living organisms. Some of the common sources of various toxic metals are listed in Table **1**.

Table 1: Sources of toxic metals in the environment

Metal	Sources
Cu	Electroplating industry, smelting and refining, mining, biosolids.
Ni	Volcanic eruptions, land fill, forest fire, bubble bursting and gas exchange in ocean, weathering of soils and geological materials, vehicle exhaust.
Pb	Mining and smelting of metalliferous ores, burning of leaded gasoline, municipal sewage, industrial wastes enriched in Pb, paints, tire wear, lubricating oil and grease.
Hg	Volcano eruptions, forest fire, emissions from industries producing caustic soda, coal, peat and wood burning.
Se	Coal mining, oil refining, combustion of fossil fuels, glass manufacturing industry, chemical synthesis (*e.g.*, varnish, pigment formulation).
Zn	Electroplating industry, smelting and refining, mining, biosolids.
Cd	Geogenic sources, anthropogenic activities, metal smelting and refining, fossil fuel burning, application of phosphate fertilizers, sewage sludge,
As	Semiconductors, petroleum refining, wood preservatives, animal feed additives, coal power, plants, herbicides, volcanoes, mining and smelting.
Cr	Electroplating industry, sludge, solid waste, tanneries, air conditioning coolants, engine parts, brake emissions.

ECO-TOXICITY OF METALS

Some of the metals like iron, cobalt, copper, manganese, molybdenum and zinc which are required by humans and other life forms for proper functioning are known as essential metals. However, at high concentrations they may have deleterious effects. Other non-essential metals like mercury, plutonium and lead are toxic metals and have no vital roles in living beings and can accumulate in the organisms over time, causing serious health hazards. Contamination of soil with heavy metals may also cause changes in the composition of soil microbial community, adversely affecting soil characteristics [4]. At high concentrations both essential and non-essential metals can damage cell membranes, alter enzyme specificity, disrupt cellular functions and damage the structure of DNA.

High concentrations of heavy metals in soil can negatively affect crop growth, as these metals interfere with metabolic functions in plants, including physiological and biochemical processes, inhibition of photosynthesis, and respiration and degeneration of main cell organelles, even leading to death of plants [5]. In humans and other higher animals they are known to affect the central nervous system (Mn, Hg, Pb, and As). Some of them are carcinogenic and some like mercury, lead, cadmium and copper have toxic effects on the kidneys or liver and nickel, cadmium, copper and chromium are known to affect skin, bones, or teeth.

TECHNOLOGIES FOR REMEDIATION OF METAL-CONTAMINATED SITES

Most of the conventional technologies for remediation of metal-contaminated sites like thermal extraction, electrokinetics, land filling *etc.* are expensive and labour intensive. Many contaminated sites across the world are left as it is without any remediation implication plan because of their economic unfeasibility. Therefore, there is an urgent need to develop innovative, eco-friendly and cost-effective technologies such as bioremediation for efficient remediation of metal-contaminated sites. Some of the technologies employed for remediation of metal-contaminated sites are listed in Table **2**.

Table 2: Technologies for remediation of metal contaminated soils

Technology	Description	Waste Contaminant	Media
Vitrification	Immobilization	Metals	Soil and sediment
Thermal extraction or thermal desorption	Volatilization at high temperature	Mercury, arsenic, cadmium	Soil and sediment
Electrokinetics	Electrical current is supplied between two electrodes, ions of contaminant will be attracted to one of the electrodes	Metals, cyanide, nitrates, and radionuclides such as uranium and strontium	Soil and sediment
Solidification/ stabilization	Immobilization of metals	Metals	Soil and sediment
In Situ Ground Water Remediation Using Colloid Technology	*In situ* colloid immobilization of contaminants	Metals absorbed on clay and silica	Ground water
Chemical oxidation	Reduction of heavy metals to lowest valence state and form stable organometallic complexes	Metals	Soil and sediment
Washing	Washing with water or surfactant or other chelators	Metals	Soil and sediment
Containment	Retention in a confined area	Metal	Soil and sediment

Table 2: contd….

| Bioremediation | Uses bacteria to transform heavy metal ions to an insoluble, less toxic form | Metals and radionuclides | Ground water, surface water, aqueous streams, Soil and sediments |
| Phytoremediation | Stabilization, extraction or volatilization of metals by plants | Metals and radionuclides | Soil and sediments, surface water |

BIOREMEDIATION AND PHYTOREMEDIATION TECHNOLOGIES

Bioremediation and phytoremediation technologies offer an eco-friendly and cost-effective approach for the remediation of metal-contaminated sites.

Bioremediation

Bioremediation is a technology that uses microorganisms and their enzymes for detoxification of pollutants. Different microbial processes can cause transformation, immobilization or solubilization of metals thereby reducing their toxic effects. Some of the bioremediation technologies involved in remediation of metal-contaminated sites include: biosorption (sequestration of metal ions on bacterial surfaces); bioaccumulation (retention and concentration of a substance by an organism); biotransformation (transformation of metals to less or non-toxic forms); and biostimulation (increase in the number and / or activity of naturally occurring microorganisms available for bioremediation by adding additional nutrients).

Phytoremediation

A subset of bioremediation which utilizes plants for the cleaning up contaminated sites is phytoremediation. This technology offers the advantages of being eco-friendly, ecologically non-disruptive, sustainable, economic, and aesthetically-pleasing as compared to other alternative technologies. Phytoremediation technologies also offer the advantage of having a wide scope and applicability; they can be applied for both organic and inorganic contaminants present in solid (soil and sludge), liquid substrates (liquid substrate) and air [6, 7]. A general overview of various phytotechnologies has been presented in Fig. **1**. In Table **3**, a summary of the mechanisms, application media, process goals, types of contaminants, plants used, and present status of various phytoremediation technologies is provided.

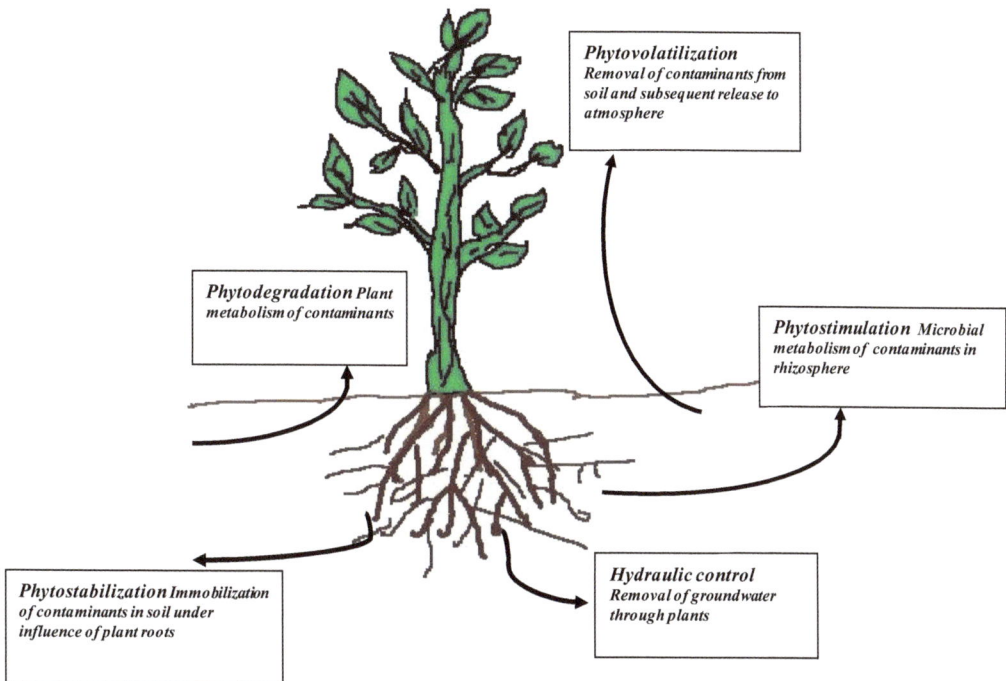

Phytovolatilization
Removal of contaminants from soil and subsequent release to atmosphere

Phytodegradation *Plant metabolism of contaminants*

Phytostimulation *Microbial metabolism of contaminants in rhizosphere*

Phytostabilization *Immobilization of contaminants in soil under influence of plant roots*

Hydraulic control *Removal of groundwater through plants*

Figure 1: Different types of phytoremediation technologies used for cleaning contaminated sites.

Table 3: Summary of application of various phytoremediation technologies (adapted from [8])

Mechanism	Process Goal	Media	Contaminants	Plants	Status
Phytoextraction	Contaminant extraction and capture	Soil, sediment, sludges	Metals: Ag, Cd, Co, Cr, Cu, Hg, Mn, Mo, Ni, Pb, Zn; Radionuclides: 90Sr, 137Cs,	Indian mustard,pennycress, alyssum, sunflowers, hybrid poplars	Laboratory, pilot, and field applications

Table 3: contd….

			239Pu, 238,234U		
Rhizofiltration	Contaminant extraction And capture	Groundwater, surface water	Metals, radionuclides	Sunflowers, Indian mustard, water hyacinth	Laboratory and pilot scale
Phytostabilization	Contaminant containment	Soil, sediment, Sludges	As, Cd, Cr, Cu, Hs, Pb,Zn	Indian mustard, hybrid poplars, poplars, grasses	Field application
Rhizodegradation	Contaminant destruction	Soil, sediment, sludges, ground water	Organic compounds (TPH, PAHs, pesticides chlorinated solvents, PCBs)	Red mulberry, grasses, hybrid poplar, cattail, rice	Field application
Phytodegradation	Contaminant destruction	Soil, sediment, sludges, groundwater surface water	Organic compounds, chlorinated solvents, phenols, herbicides, munitions	Algae, stonewort, hybrid poplar, black willow, bald cypress	Field demonstration
Phytovolatilization	Contaminant extraction from media and release to air	Groundwater, soil, sediment, sludges	Chlorinated solvents, some inorganics (Se, Hg, and As)	Poplars, alfalfa black locust, Indian mustard	Laboratory and field application
Hydraulic control (plume control)	Contaminant degradation or containment	Groundwater, surface water	Water-soluble organics and inorganics	Hybrid poplar, cottonwood, willow	Field application
Vegetative cover (evapotranspiration cover)	Containment, erosion control	Soil, sludge, sediments	Organic and inorganic compounds	Poplars, grasses	Field application
Riparian corridors (non-point source control)	Contaminant destruction	Surface water, groundwater	Water-soluble organics	Poplars	Field demonstration

PHYTOREMEDIATION TECHNOLOGIES FOR RECLAMATION OF HEAVY METAL-POLLUTED SOIL AND WATER

Different plants can grow in metal-contaminated sites but all of them may not be suitable for phytoremediation of metals. A plant should possess the following characteristics to be used for metal phytoremediation: 1) tolerance to metals, 2) ability to accumulate the metal to the above-ground parts (for phytoextraction) and below-ground parts (for phytostabilization and landscape recreation); 3) fast growth with high biomass production, 4) high metal accumulation potential; and 5) easily harvestable.

Phytovolatilization

During phytovolatilization, the soluble contaminants are taken up with water by the roots, transported to the leaves, and volatilized into the atmosphere through the stomata. Volatilization of many metals like selenium, arsenic and mercury by plants has been reported. Inorganic Se is taken up by plants and assimilated to organic selenoaminoacids, selenocysteine, and selenomethionine. The latter can be biomethylated to form dimethylselenide which is volatile and is lost to the atmosphere. *Brassica juncea* is an important plant for Se removal from soils [9]. Rhizosphere bacteria are known to play an important role in Se volatilization by increasing accumulation potential and consequently volatilization of Se by the plants [10]. Apart from Se, Hg and As are also susceptible to volatilization by plants. Mercury occurs in the environment mainly as divalent cations Hg^{2+} which are not easily volatilized. However, it may be reduced to elemental mercury by rhizobacteria thus increasing the volatilization abilities of the associated plants. Successful volatilization of mercury has been reported by genetic manipulation in plants. Bacterial mercury reductase (mer A) gene was incorporated in *Arabidopsis* and yellow poplar plants. The transgenic plants could convert Hg^{2+} to elemental forms [11, 12], and subsequently volatilize them. Phytovolatilization of As has also been documented but it is of meager importance. In contrast to the phytovolatilization of selenium and mercury which are converted to nontoxic methylated Se and elemental mercury, respectively, arsenic is volatilized to arsines which are toxic in nature. Therefore, care has to be taken during any of the phytoremediation technologies that the metals are not released in toxic volatile forms.

Phytostabilization

Phytostabilization is a mechanism that immobilizes contaminants —mainly metals—within the root zone, thus limiting their migration. Immobilization of contaminants can result from adsorption of metals to plant roots, formation of metal complexes, precipitation of metal ions (*e.g.*, due to a change in pH), or a change to a less toxic redox state which causes a decrease in the mobility and bioavailability of metals. Another similar phenomenon called **phytosequestration** may also occur, where the contaminants are sequestered within the plant cells, particularly in vacuoles, as a storage and waste receptacle for the plants. During phytosequestration contaminants are generally retained in the roots and are not translocated further. Thus it does not require plant harvesting and disposal [13]. However, evaluation of the system is necessary to verify that translocation of contaminants into plant tissues other than the roots is not occurring. Due to the continuing presence of contaminants in the root zone, plant health must be monitored and maintained to ensure system integrity and prevent future release of contaminants.

Phytostabilization and phytosequestration offer the advantage that the contaminants are immobilized in soil and thus the harvesting of plants is not required. These technologies can be used to prevent migration of soil contaminants with wind and water erosion, soil dispersion, and leaching.

Phytoextraction (Phytoaccumulation)

As mentioned before, phytoextraction may be defined as the extraction of metals from soil or sediment by metal-accumulating vascular plants. These plants take up the metals from the external medium and translocate them to the roots and shoots. The metals are ranslocated to the above-ground parts of the plants through vascular tissues and are compartmentalized in vacoules and cell wall in different organs. The above-ground biomass of plants is harvested and metals can either be recycled (metals of economic importance) or disposed off (harmful metals and radionuclides). It is also called **phytoaccumulation** as the inorganic contaminants are accumulated in different parts of the plant. Some plant species which can accumulate high concentrations of metals are called hyperaccumulators.

Phytoextraction of many metals like cadmium (Cd), lead (Pb), copper (Cu), zinc (Zn), and nickel (Ni) has been demonstrated [14]. Phytoextraction of As by hyperaccumulator plants and plants with high biomass such as ferns *Ptris vittata* and *Ptris calomelanos* have also been reported [15]. Several plant species including *Thlaspi caerulescens* have been shown to accumulate very high levels of Zn and Cd from soils. *Brassica juncea* has also been found to be an excellent accumulator of metals like Cd, Cr, Ni, Zn, and Cu [6, 16] and several plant species have been shown to accumulate Pb [17]. Plants, effective in removing metals from aquatic systems include *Eichhornia crassopes*, *Hydrocotyle umbellata*, *Lemna minor*, *Scirpus lacustris*, *Bacopa monnieri*, and *Azolla pinnata* [18].

Some of the obstacles for commercial implementation of phytoextraction include the disposal of contaminated plant materials, slow plant growth, insufficient biomass production in plants, and shallow root systems. Low metal bioavialabilty and poor translocation of metals from roots to shoots often limit the efficiency of the technology. Still it is one of the best methods for removing contaminants from soil without disturbing the ecosystem.

However, there have been several approaches for enhancement of naturally occuring phytoextraction including chelate-assisted phytoextraction or induced phytoextraction, during which artificial chelates are added to soils to facilitate the mobility and uptake of metal contaminants. Increase in the amount of bioavailable metals in soil and higher levels of accumulation in plants in the presence of chelating agents like ethylene diamine tetra acetic acid in Pb-contaminated soils have been reported [19].

Rhizofiltration

The roots or root exudates may create conditions that result in precipitation of contaminants on the roots. The contaminant may remain associated with the roots or may be taken up and translocated to other parts of the plant, depending on the plant species, type of contaminant and its concentration. Rhizofiltration can be applied for decontamination of large volumes of water with low contaminant concentrations (in the ppb range). It has primarily been applied for the removal of metals (Pb, Cd, Cu, Fe, Ni, Mn, Zn, Cr(VI) [6] and radionuclides of 90Sr, 137Cs, 238U, 236U [17]. Mechanism of uptake of metals by plant roots and fate of metals in plants are presented in Fig. **2**.

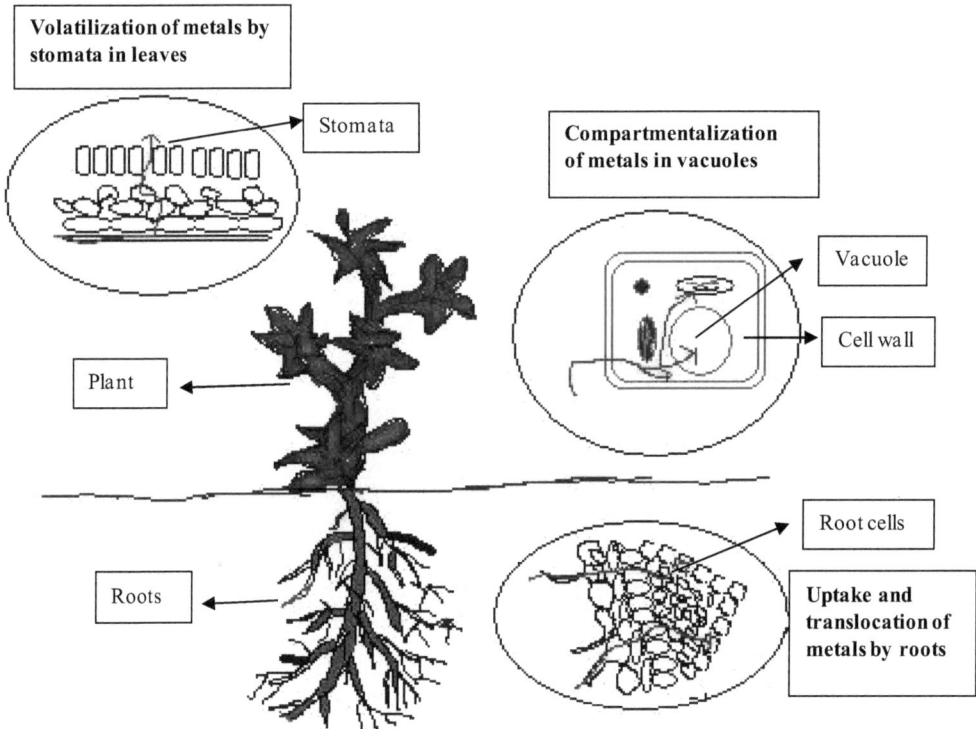

Figure 2: Fate of metals in plants during various phytotechnologies (Adapted from [21]).

Rhizofiltration can be conducted *in situ* to remediate contaminated surface water bodies, or *ex situ*, in which an engineered system of tanks can be used to hold the introduced contaminated water and the plants. Once the inorganic toxic pollutants have been accumulated in the plants proper disposal of the contaminated plant biomass is required. Applications of rhizofiltration are currently at the pilot-scale stage. Phytotech tested a pilot-scale rhizofiltration system in a greenhouse at a Department of Energy uranium-processing facility in Ashtabula, Ohio [17, 20]. This engineered *ex situ* system used sunflowers to remove uranium from contaminated ground water and/or process water. Phytotech also conducted a small-scale field test of rhizofiltration to remove radionuclides from a small pond near the Chernobyl reactor, Ukraine. Sunflowers were grown for four to eight weeks in a floating raft on a pond, and bioaccumulation results indicated that sunflowers could remove 137Cs and 90Sr from the pond.

HEAVY METAL DETOXIFICATION MECHANISMS IN PLANTS

There are different mecahnisms by which plants detoxify metals. One of the mechanisms of heavy metal tolerance in plants is control of cytosolic concentration of heavy metals by active plant efflux systems. Additionally, plants chelate metal ions extracellularly by components of root exudates and/or bind them to the rhizodermal cell walls and subsequently take up the metals. Intracellularly, plant cells produce chelating agents such as phytochelatins and metallothioneins [22, 23], which have high-affinity metal binding properties. The resulting complexes can finally be exported from the cytoplasm across the tonoplast and become sequestered inside the vacuole [24]. Also other organelles are involved in metal storage. For example, Fe is stored bound to ferritin inside chloroplasts [25].

Once inside the plant, most metals are too insoluble to move freely in the vascular system, so they usually form carbonate, sulfate or phosphate precipitates or may also be chelated in the cytosol to metallothioneins, organic acids, amino acids and metal specific chaperons. Thus, the metal complexes are immobilized in apoplastic (extracellular) and symplastic (intracellular) compartments. From both comprtments they enter the xylem. The transport of ions into the xylem is membrame transport based on protein-mediated process and is highly controlled. Metals are then compartmentalized into organelles like vacoules or cell wall, to decrease the potential damage to the cytosol.

INTERACTIONS BETWEEN MICRORGANISMS AND PLANTS

Plant-microbe interactions form the basis of many important biological phenomena such as biological nitrogen fixation whereby bacteria from the genus like *Rhizobium* and others fix atmospheric nitrogen in the root nodules of leguminous plants. This phenomenon is possible only by the involvement of both the organisms. Leguminous plants have evolved together with rhizobia to contribute heavily to the successful symbiosis by expressing more than 20 bacterial genes specifically for this purpose. The genes responsible for nitrogen fixation, the *nif* genes, are present in bacteria and the plants provide protection, food and shelter to the bacteria in addition to protection to the bacterial enzyme nitrogenase which is highly sensitive to oxygen by producing a protective protein leghemoglobin. Yet another classical example of plant-microbe interaction is that of *Agrobacterium tumefaciens*, which causes the plant disease crown gall in dicotyledonous plants, characterized by the formation of tumours or crown galls.

There are innumerable interactions going on between the plants and the microbes, most of which occur underground in the rhizosphere. Many microorganisms inhabiting the rhizosphere live in symbiotic relationships with plants and possess properties by virtue of which they help in plant growth and development and thus are termed as plant growth promoting microorganisms (PGPM). Interactions between plants and the rhizosphere microorganisms can be positive, negative or neutral. Some of the common examples of positive interactions include symbiotic nitrogen fixation, mycorrhizal fungus, biocontrol agents and plant growth promoting bacteria. Negative interactions include pathogenesis by bacteria or fungi. Plant-microbe interactions can positively influence plant growth through a variety of mechanisms, including fixation of atmospheric nitrogen by different classes of proteobacteria, increased biotic and abiotic stress tolerance imparted by the presence of endophytic microbes, and direct and indirect advantages imparted by rhizosphere microorganisms. Bacteria can also positively interact with plants by producing protective biofilms or antibiotics and thus operating as biocontrol agents against potential pathogens, or by degrading plant- and microbe-produced compounds in the soils that would otherwise be allelopathic or even autotoxic. Since, most of the beneficial interactions between plants and microorganisms occur underground through the roots, it is essential to understand the micro-environment of 'rhizosphere'.

The Rhizosphere

The term rhizosphere refers to the environment influenced by the roots of vascular plants in which elevated bacterial activity is observed [26]. In general, the rhizosphere has been described as the zone of soil under the direct influence of plant roots, which usually extends a few millimeters from the root surface and is a dynamic environment for microorganisms where complex biological and ecological processes occur. This complex and dynamic microenvironment where bacteria, fungi and actinomycetes along with other microorganisms, in association with roots forms unique communities and has considerable potential for detoxification of hazardous organic and inorganic pollutants. Some of the common microorganisms found in the rhizosphere are listed in Table **4**.

Table 4: Common bacterial genera found in soil

Bacteria	Function
Azospirillum	N-fixation in tropical grasses
Pseudomonaceae	Decompose organic matter
Azotobacteriaceae	Fee N fixers
Rhizobium	Symbiotic N fixers, produce root nodules in legumes
Agarobacterium	Produces root infections
Enterobacter (Aerobacter)	Decomposition of organic matter
Nitrobacteriaceae	Oxidize reduced forms of inorganic Nitrogen
Bacillus	Some sp. N-fixers, produce a number of extracellular enzymes which hydrolyse complex organic material
Clostridium	Ferment cellulose, starch, pectin and sugars
Actinomycetes	Largest and heterogenous group
Frankia	Fix nitrogen in non legumes
Arthobacter *Nocardia* *Actinomyces* *Streptomyces*	Use variety of organic compounds and produce antibiotics

A majority of the microbial population in the soils is associated with the plant roots due to the availability of high levels of nutrients (especially small molecules such as amino acids, sugars and organic acids) that are exuded from the roots of most plants, and thus are termed root exudates. Root exudates can stimulate microbial growth in the immediate vicinity of the roots (rhizosphere) causing in high abundances of microbes [27] which fulfill important ecosystem functions for plants and soils.

Root Exudates and their Role in Phytoremediation

In addition to accumulating biologically active chemicals, plant roots continuously produce and secrete into the rhizosphere compounds including ions, free oxygen and water, enzymes, mucilage, and a diverse array of carbon-containing primary and secondary metabolites.

Root exudation can be broadly divided into two active processes:

- **Root excretion:** It involves gradient-dependent output of waste materials with unknown functions.

- **Secretion:** involves exudation of compounds with known functions, such as lubrication and defense.

Roots Release Compounds Via at Least two Potential Mechanisms:

Root exudates are transported across the cellular membrane and secreted into the surrounding rhizosphere. Plant products are also released from root border cells and root border-like cells, which separate from roots as they grow.

Root Exudates are Often Divided Into:-

Low-molecular weight compounds such as amino acids, organic acids, sugars, phenolics, and other secondary metabolites account for much of the diversity of root exudates.

High molecular weight exudates such as mucilage (polysaccharides) and proteins are less diverse but often compose a larger proportion of the root exudates by mass.

Although the functions of most root exudates have not been determined, several compounds present in root exudates play important roles in biological processes. Root exudates play a significant role in phytoremediation technologies because they are responsible for creating a high density zone of active and diverse heterotrophic microorganisms on root surfaces and the rhizosphere. This situation often facilitates selection of metal-resistant microorganisms and horizontal transfer of resistant genes may take place [28]. Though a diverse group of chemical compounds are secreted from the roots in the form of root exudates, carboxylic acids and phenolic acids play an important role in solubilization of metals. Excessive release of these organic acids causes decline in the pH and thus increases the solubility and availability of unavailable forms of metals which subsequently may be taken up by the hyperaccumulator plants. Not only organic acids but polysaccharides secreted from the plant roots can also contribute to phytoremediation. Polysaccharides can bind to and immobilize metal ions into the soil thereby restricting their uptake by the plants, thus helping in phytostabilization of metals [29].

Plants may also secrete metal chelators, phytochelatins and phytosiderophores, and organic acids into the rhizosphere. Consequently, the availability of metallic soil micronutrients including iron, manganese, copper, and zinc could be increased. Metal chelators form complexes with soil metal ions, thus releasing metals that are bound to soil particles. This would lead to an increase in metal solubility and thereby facilitate phytoextraction.

Beneficial Bacteria in Rhizosphere

Beneficial free-living soil bacteria are generally referred to as plant growth-promoting rhizobacteria (PGPR) and are found in association with the roots of many different plants. These bacteria can positively influence plant growth and development indirectly or directly [30, 31]. Indirect mechanisms used by bacteria include protection against pathogenic bacteria by antibiotic production, reduction of iron available to phytopathogens in the rhizosphere, synthesis of fungal cell wall-lysing enzymes, and competition with detrimental microorganisms for sites on plant roots. Direct mechanisms of plant growth by bacteria include an increase in phosphorus availability for plants, nitrogen fixation for plant use, sequestration of iron for plants by siderophores, production of plant hormones like auxins, cytokinins and gibberellins, and lowering of plant ethylene levels [31]. In addition, these populations are known to affect heavy metal mobility and availability to the plant through release of chelating agents, acidification, phosphate solubilization, and redox changes [32, 33]. A particular bacterium may affect plant growth and development using any one, or more, of these mechanisms and a bacterium may utilize different mechanisms under different conditions.

Some of the common mechanisms by which soil bacteria promote the growth and development of plants are depicted in Fig. **3**.

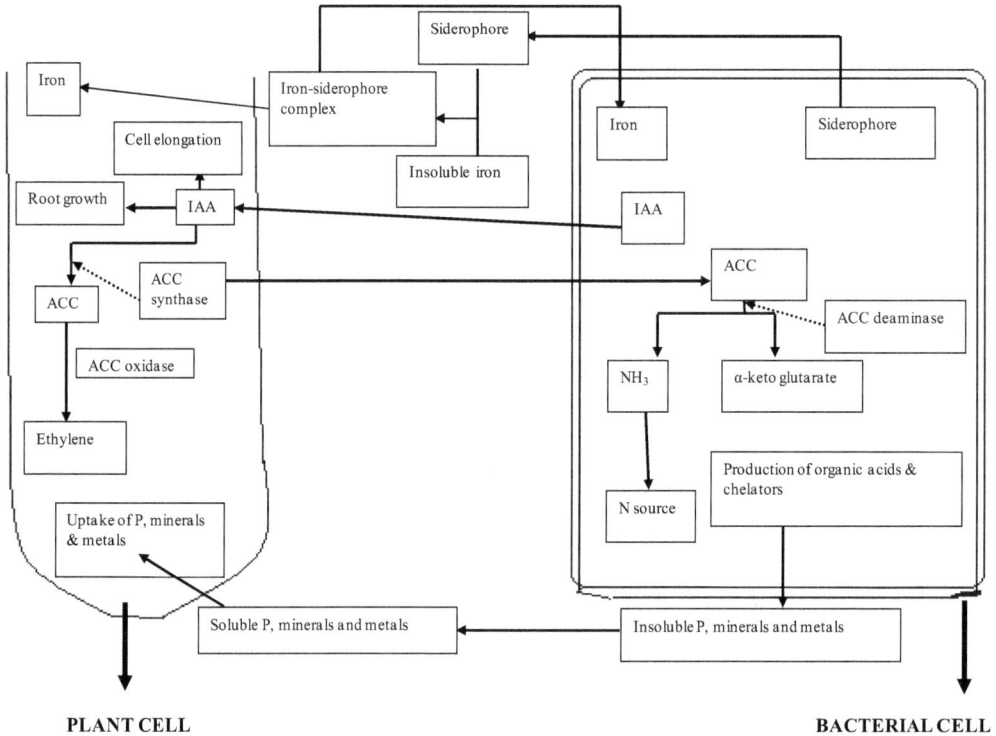

Figure 3: Mechanisms by which rhizosphere bacteria promote plant growth. Dotted arrows represent enzyme activities, ACC= 1-aminocyclopropane-1-carboxylate, IAA= indole acetic acid (Adapted from [34].

These microorganisms in soils directly or indirectly contribute to plant growth and hence to phytoremediation practices. An important approach for successful phytoremediation is to enhance metal accumulation and high biomass production in plants growing in metal- contaminated soils. Inoculation with beneficial microorganisms not only might increase metal uptake by plant roots but it could certainly help in achieving large above-ground plant biomass. Thus a much higher metal removal would be possible in bacteria-inoculated plants as compared to those growing in their absence.

Various mechanisms by which soil beneficial bacteria facilitate plant growth and thus phytoremediation are discussed as follows:

The Role/Use of Rhizobacteria in Phytoremediation

Some of the important mechanisms by which bacteria contribute to growth and develpoment of plants in metal-contaimated soils and therefore to phytoremediation are listed as follows:

Root Development

Plant growth promoting bacteria are known to improve plant growth and development in heavy metal-contaminated soils by assisting root growth and branching. Two strains of rhizobacteria *P. fluorescens* (92rk and P190r) when inoculated in tomato and cucumber roots were found to increase total length, surface area, and volume in the roots [35]. Furthermore, production of phytohormones like indole acetic acid also results in an increase in growth and development of the plant roots. Bacteria also help in root development by inducing resistance against root pathogens. *P. fluorescens* has induced systemic resistance and inhibited early root penetration of Heterodera schachtii, the cyst nematode in sugar beet [36]. Thus, these bacteria by various mechanisms modify the root architecture and favour phytoremediation

Effect on Plant Growth

Lowering of Plant Ethylene Levels by Production of ACC Deaminase

Plants respond to heavy metal stress by producing high levels of a phytohormone, ethylene. Ethylene plays a critical role in various developmental processes, such as leaf senescence, leaf abscission, epinasty and fruit ripening *etc*. However, at higher concentrations ethylene inhibits plant growth and development [37-39]. It has been demonstrated that many rhizobacteria synthesize 1-aminocyclopropane-1-carboxylate (ACC) deaminase, which metabolizes ACC of the roots to ketobutyrate and ammonia [40]. The bacteria utilize the NH_3 evolved from ACC as a source of nitrogen and thereby restrict the accumulation of ethylene within the plant, which otherwise inhibits plant growth. In this way many bacteria promote plant proliferation through the production of ACC deaminase.

Siderophore Production

Siderophores are organic molecules having high affinity for Fe(III) ions, but can also chelate other metal ions [41, 42]. Many soil bacteria are known to produce siderophores which regulate the availability of Fe in the plant rhizosphere. It has also been found that competition for iron in the rhizosphere is controlled by the affinity of the siderophores for iron. In soils, siderophore producing bacteria make the iron soluble from insoluble forms thus making it avialable to plants and contributing to plant growth in iron-deficient conditions. Furthermore, siderophores may also play an important role for the mobilization of metals in the rhizosphere. Experiments with *Brassica juncea* revealed that the inoculation of siderophore-producing bacteria, *Pseudomonas* sp. or *Bacillus megaterium*, significantly increased the growth of plants in Ni-contaminated soil without showing any symptoms of toxicity as compared to the controls [43, 44]. It has been reported that 83% bacterial isolates recovered from within *A. bertolonii* produced siderophores and many of them were found to help in plant growth and Ni hyperaccumulation [45]. Similarly, siderophore production in Ni-resistant bacteria isolated from *T. goesingense* has been reported [46].

Phosphorus Solubilization

In soils with high metal concentrations metal-resistant phosphorus solubilizers may be of importance as they offer a biological rescue system capable of solubilizing the insoluble inorganic P in soils. This makes more phosphorus available to the plants and thus helps the plant to survive under unfavourable conditions. Basic mechanism of phosphorus solubilization by bacteria is by production of organic acids which create an acidic environment thus solubilizing the unavialable forms of phosphorus [47]. However, these acids can also bind to metal ions by their carboxylic group and thus increase the bioavialability of metals, and facilitate metal uptake by hyperaccumulators.

Indole-3-Acetic Acid Production

The production of indole-3-acetic acid, a phytohormone of auxin series which acts as a plant growth promoter, by rhizosphere bacteria is yet another characteristic believed to play an important role in plant– bacterial interactions and plant growth in metal-contaminated soils [43, 44]. A low level of IAA produced by rhizosphere bacteria promotes primary root elongation whereas a high level of IAA stimulates lateral and adventitious root formation. Thus, these bacteria can facilitate plant growth by altering the plant hormonal balance. A metabolic precursor of IAA, anthranilic acid, can reductively solubilize soil Fe (III), and increase its availability *via* a mechanism different from that involving siderophores [48]. In a study, addition of IAA to soil enhanced the uptake of iron and other elements (*e.g.,* zinc, calcium *etc.*) in plant roots [49, 50].

Resistance Against Plant Pathogens

Another effect by which bacteria influences plant growth positively is by inducing resistance in plants against fungal, bacterial and viral diseases, and insects including nematode pests [51]. The induction of systemic resistance in plants by rhizobacteria is referred to as Induced Systemic Resistance (ISR). Antibiotic-secreting plant growth-promoting bacterial strains can inhibit the proliferation and subsequent invasion of phytopathogens, hence protecting plants from the effect of pathogens [52]. Experiments showed that seed treated with *P. fluorescens* strain 97 protected beans against halo blight disease caused by Pseudomonas syringae pv. phaseolicola [53].

Additionally, they provide different mechanisms for suppressing plant pathogens. They include competition for nutrients and space, antibiosis by producing antibiotics *viz.,* pyrrolnitrin, pyocyanine, 2,4-diacetyl phloroglucinol and production of siderophores (fluorescent yellow green pigment), *viz.,* pseudobactin which limits the availability of iron necessary for the growth of pathogens. Other important mechanisms include production of lytic enzymes such as chitinases and β-1,3-glucanases which degrade chitin and glucan present in the cell wall of pathogenic fungi, HCN production and degradation of toxin produced by pathogen.

Bioavailability of Heavy Metals

Soil microorganisms can differently influence the speciation of metals in soils. Consequently, altering soil metal bioavailability might be altered and affect the efficiency of phytoremediation. Bacteria in the rhizosphere may transform metals by altering the chemical properties of the environment such as pH, organic matter content, redox state, *etc.* to forms which are readily taken up into the roots. In a study, the rhizobacteria were found to enhance Ni uptake by *Alyssum murale* by releasing the metal from non-soluble phase probabaly as a result of a decrease in the pH of the environment or production of siderophores [32]. Soil bacteria were also found to enhance Se accumulation in plants by reducing selenate to organic Se, and organoselenium forms like SeMet which are known to be taken up at faster rates into the roots than inorganic forms [51]. The relative changes of organic-bound Cu, Zn and Pb were, respectively, +5%, +23%, +3% in the rhizosphere with PGPR, and 0.8%, −3%, −2% in the control. Thus, significant amounts of Cu, Zn and Pb were bounded by organic matter in the rhizosphere with PGPR. Similarly, a strain of *Pseudomonas maltophilia* was found to diminish the mobile and toxic Cr^{6+} to nontoxic and immobile Cr^{3+}, and also to minimize environmental mobility of other toxic ions such as Hg^{2+}, Pb^{2+}, and Cd^{2+} [54, 55].

Interactions Between Rhizobacteria and Plants

Contaminated soils are often nutrient poor or sometimes nutrient deficient due to the loss of beneficial microbes. This result in poor plant growth and reduced plant biomass. However, such soils can be made nutrient rich by applying metal-tolerant microbes, especially the plant growth promoting rhizobacteria which would provide not only the essential nutrients to the plants growing in the contaminated sites but would also play a major role in detoxifying heavy metals to assist phytoremediation of heavy metal-contaminated soils. Microbial community structure of the rhizopshere is important for the establishment and growth of plants. Microbial populations often undergo some kind of positive relationships with the plants under metal stress conditions so that both the partners can survive under such unfavorable conditions. A diagrammatic representation of some of the interactions between soil, microorganisms and plants is depicted in Fig. **4**.

Many studies have indicated that inoculation of the soil with plant growth promoting microorganisms results in better phytoremediation efficiency. It has also been reported that a high population of metal-resistant bacteria exist in the rhizosphere of hyperaccumulators *Thlaspi caerulescens* [56] and *Alyssum bertolonii* [57] or *Alyssum murale* [32] grown in soils contaminated with Zn and Ni, respectively. The presence of rhizobacteria also causes in increased accumulation of Zn [58], Ni [31] and Se [10] in *T. caerulescens*, *A. murale* and *B. juncea*, respectively.

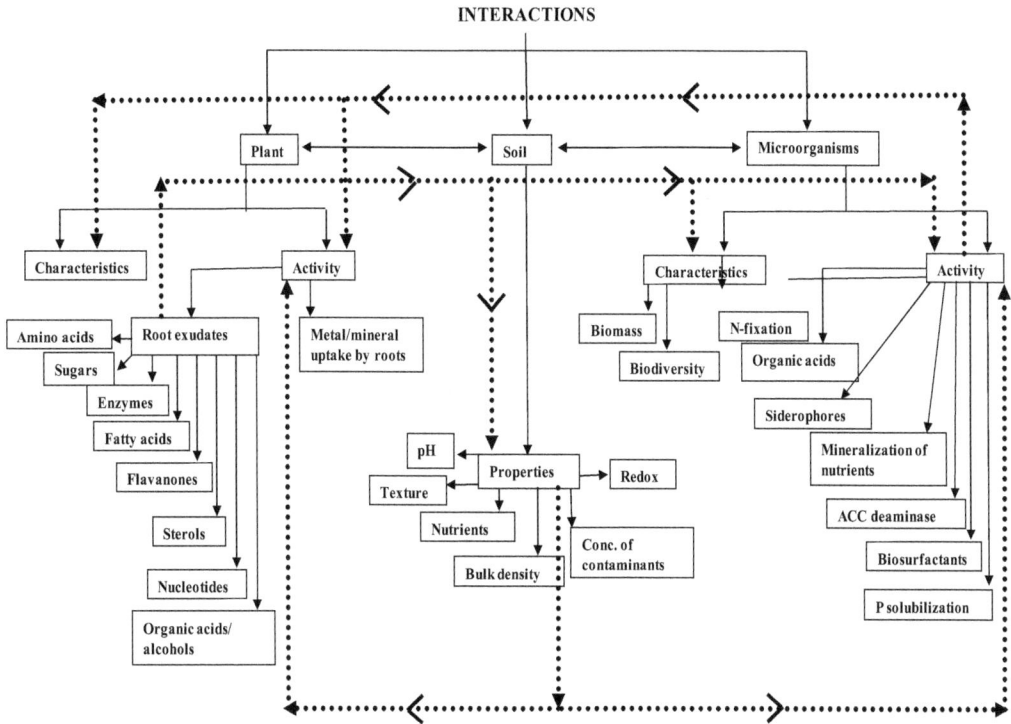

Figure 4: Interactions between plants, microorganisms and soil (dashed arrows represent interactions).

Bacteria existing in soils with prolonged heavy metal exposure often develop resistance towards them and multi-metal resistance property is common in rhizobacterial isolates. Patterns of tolerance of the heavy metals in the 107 rhizobacterial isolates was tested at 1 mmol/L [59] and it was that found all the rhizobacterial strains were tolerant to multiple metal ions. Strains with hexa-, penta-, tetra-, and tri-metal ions tolerant, respectively, were found more frequent than those with hepta-, double and mono-tolerance. Notably cadmium, copper, lead, and nickel resistance seemed to be restricted to those strains which were resistant to six metals or more.

Soil bacteria have also been reported to play an important role in phytovolatilization of selenium. It was evidenced that bacteria facilitate the uptake of selenium by plants thereby increasing its volatilization. Similarly, in presence of available phosphorus, phosphate transporters are activated which also take up arsenate [60].

Many of the soil microorganisms are known to exude some secretions such as antibiotics (including the antifungals), carboxylic acid, hydrocyanic acid, indoleacetic acid (IAA), siderophores, and 1-aminocyclopropane-1-carboxylic acid (ACC) deaminase. These substances could increase bioavailability and facilitate root absorption of heavy metals such as Fe and Mn as well as non-essential metals such as Cd, and enhance tolerance of host plants by improving the P absorption and promote plant growth.

It has also been reported that presence of beneficial microbes in the rhizosphere lowers the toxicity of heavy metals such as Ni, Zn, Pb, Cr, Cd, As to plants and thereby aids plant growth. Microbial population in the rhizosphere is known to influence heavy metals uptake by plants. It is reported that under a non-sterile soil system, plants showed no iron-deficiency symptoms and had a fairly high iron level in the roots compared to plants grown in a sterile system. This can be attributed to rhizospheric microbial activity which plays an important role in iron acquisition. The correlation between metal resistance and metal mobilization abilities of rhizobacteria was investigated under heavy metal stress [59]. The highest incidence of the PGPR property of isolates and metal resistance were recorded for: phosphate solubilizers with Cr, Zn and Pb (92.5%, 82.2% and 68.2%), respectively; then for siderophore producers with Cr, Zn and Pb (78.5%, 71.02% and 61.6%), respectively, and finally for acid producers with Cr, Zn and Pb (63.5%, 53.3% and 42.9%), respectively. Thus, rhizosphere bacteria have adopted different mechanisms for metal tolerance.

Rhizobacteria may also stimulate uptake of metals by plants by various mechanisms. For example, bacteria may stimulate the sulfate transport protein, localized in the root plasma membrane, which also transports selenate [61]. Inorganic Hg uptake in higher plants is not well investigated, but has been linked to the passive uptake of lipophilic chloride complexes in phytoplanktons [62].

It has also been evidenced that the chemical conditions of the rhizosphere differ from those of the bulk soil as a consequence of various processes that are induced by the plant roots and/or by the rhizobacteria. Plant-bacteria interactions could stimulate the production of compounds that could alter soil chemical properties in the rhziosphere and enhance heavy metal accumulation in plants. For example, soil acidification in the rhizosphere of *Thalaspi caerulescens* facilitates metal ion uptake by increasing metal ion mobility around the roots [56]. It was also reported that the accumulation of Hg increases when the pH of the culture solution was lowered. It was hypothesized that rhziobacteria of the plants reduced the pH in the rhizosphere and thereby increased Hg uptake into plants [10]. A further study examined the influence of hydrogen and aluminum ions on the growth of the nitrogen-fixing and growth-promoting bacteria including *Azospirillum lipoferum* 137, Arthrobacter mysorens 7, *Agrobacterium radiobacter* 10, and Flavobacterium sp. L30. It was found that the response of plants to the inoculation strongly varied with the soil pH.

In addition, many rhizobacteria can remove a number of metals from the environment by reducing them to a lower redox state [63]. Many of the microorganisms can use the metals as terminal electron acceptors in anaerobic respiration. Most of these dissimilatory metal-reducing bacteria use Fe^{3+} and S^0 as terminal electron acceptors [63]. The microbes can also reduce Cr^{6+} to Cr^{3+}, either aerobically or anaerobically [64]. Such microbial transformations are of importance in phytotechnologies like phytostabilization as they assist the plants in immobilization of metal contaminants and many times they convert Cr to less or non toxic forms.

As discussed previously, the efficacy of phytoremediation is often increased by using chelators such as EDTA and organic acids which enhance the availability of meals for plant uptake. Instead of using synthetic chelators, a better approach is use of natural chelators directly or application of bacteria producing these chelators at the metal-contaminated site. Some of the common chelating molecules which have immense scope for application for phytoremediation and are commonly secreted by microorganisms are organic acids, siderophores and biosurfactants. Rhizosphere bacteria are known to facilitate phytoaccumulation of toxic metals by increasing the availability of these metals mainly by production of these metal-chelating molecules.

The benefits of siderophore-producing bacteria (SPB) for removal of metals from contaminated soils using plants have been demonstrated. Metal-resistant siderophore producers play an important role in the successful survival and growth of plants in contaminated soils by alleviating metal toxicity and supplying the plants with nutrients, particularly iron. Furthermore, bacterial siderophores are able to bind metals other than iron and thus enhance their bioavailability in the rhizosphere of plants [34].

Biosurfactants are yet another group of biomolecules which are surface active agents produced by living organisms like microbes and plants. In general, they possess a polar head and non polar fatty acid chain. They are known to chelate divalent, trivalent, metal cations as well as metal anions [65, 66] *via* various binding groups like carboxylic acid group in rhamnolipid. When applied to soil, they reduce the interfacial tension, thus causing increase in the metal mobility and availability. In some recent studies potential of bacterial biosurfactants for enhanced accumulation of metals by plants have been reported. The biosurfactant-producing and heavy metal-resistant *Bacillus* strain J119 could colonize the rhizosphere soils of maize, Sudan grass, and tomato [67]. Application of the strain was found to increase biomass and Cd uptake.

Thus, the rhizobacteria capable of producing such metal chelators are promising candidates for chelate-enhanced phytoremediation as chemical chelators like EDTA are expensive and cause environmental pollution. Additionally, rhizobacteria help in plant growth and development in metal-contaminated sites by secreting growth promoting compounds. They are also directly involved in reduction of toxicity, change in speciation, increase or decrease mobility of the metals. Many rhizobacteria produce extracellular polysaccharides which reduce the bioavailability of metals in soil as well [29]. There are examples of bacteria which can modify their surfaces under high metal concentrations to increase their survival rate under toxic conditions.

Role of Mycorrhizal Associations in Phytoremediation of Metals

Among the microorganisms that affect rhizosphere processes, symbiotic fungi forming mycorrhizae are of extreme significance as they are capable of inducing a series of changes in plant physiology, nutrient availability and microbial composition that may determine the efficiency of a phytoremediation attempt. Beyond the rhizosphere, mycorrhizal hyphae act as the roots of the roots, and may thus extend the rhizosphere into the bulk soil by creating a new interface of soil-plant interactions: the hyphosphere.

Mycorrhiza can be defined as association between fungi (myco) and plant roots (rhiza). Studies with mycorrhizal fungus have focused on their ability to provide the host plant with nutrients in nutrient deficient soil [68] and have ignored their other roles in a soil ecosystem. Although mycorrhizal fungi are well known as saprophytic microorganisms, there is a possibility that they may retain the ability to degrade a large assortment of compounds available in the soils including organic pollutants. Mycorrhizal fungi are usually classified as belonging to one of four major types: the ectomycorrhizae, the vesicular arbuscular mycorrhizae, the ericaceous mycorrhizae, and the orchid mycorrhizae.

An important benefit of mycorrhizal infection has been observed in sites with heavy metal contamination including mine spoils and at sites containing high concentrations of toxic metals. Plants with mycorrhizal associations appear to be protected from the phytotoxic effect of heavy metals. It is believed that the mycorrhizal fungi present in the contaminated sites have been selected over time with greater resistance to heavy metals. Heavy metals are believed to be bound by carboxyl groups in the hemicelluloses of the interfacial matrices between the host cells and the fungi.

Study of the roles of mycorrhiza in metal speciation in the rhizosphere and the impact on increasing host plant tolerance against excessive heavy metals in soils showed that speciations of Cu, Zn and Pb changed significantly in the rhizosphere of AM (arbuscular mycorrhiza) infected and non-infected maize. The greatest change was exchangeable Cu that increased by 17% in AM infected rhizosphere as compared to the non-infected rhizosphere. The results showed that mycorrhiza had a protective role for its host plants from the phytotoxicity of high concentrations of metals like copper, zinc and lead which was probably due to the mycorrhizal effect on the speciation of metals.

It was observed that accumulation of copper and zinc and other metals in the roots and shoots of mycorrhiza-associated plants was significantly lower than those in the plants without AMF [69, 70]. This suggested that mycorrhizae efficiently restricted excessive copper and zinc absorption into the host plants. Another mechanism by which AMF helps in phytoremediation is by changing the chemical properties of metals in the rhizosphere and enhancing metal accumulation by plants. AMF are known to produce metallothioneins and cysteine-rich proteins which facilitate metal accumulation by plants.

It has also been suggested that AMF association enhances the translocation and thus sequestration of toxic metals within plants [70]. Furthermore, enhancement in the transcription of phytochelatin synthetase and metallothionein genes was found in the presence of mycorrhizal associations in certain plants [70]. AMF also improve plant growth and development in heavy meta-contaminated soils by assisting root growth and branching. The total root length, surface area, and volume in tomato and cucumber roots increased with AMF *Glomus mosseae BEG12* [35]. The plants developed increasedroot surface area, volume, number of tips and degree of root branching. This development of root structure aids in scavenging of phosphorus and other nutrients by the plants.

Genetic Engineering Approach for Phytoremediation

Recent advances in the study of plant-microbe interactions have enabled the construction of plants that are even more selective for certain microorganisms in the rhizosphere. Many different bacteria communicate *via* chemical signals including amino acids or short peptides (gram-positive bacteria) and fatty acid derivatives (gram-negative bacteria). Communication by gram-negative bacteria *via* *N*-acyl-homoserine lactones (AHL) is one of the most characterized systems [71]. AHL is produced by bacteria at low levels when cell densities are low. When cell numbers increase sufficiently, enough AHL is produced by the community to interact with a transcriptional regulator. The AHL–regulator complex can then bind to target promoter sequences thus initiating gene transcription. This communication system allows the microorganisms to orchestrate a concerted response to a stimulus. Many different microorganisms also interact either positively or negatively due to cross-communication from the various chemical signals including AHLs. Additionally, it has been shown that some plants produce chemicals that can mimic the bacterial signals.

It has been demonstrated that plants engineered with *N*-acyl-homoserine lactones (AHL) genes showed increased resistance against fungal disease because it contributes to antifungal activity of the biocontrol agent *Pseudomonas aureofaciens* into a tobacco plant [72]. Similarly, an opine-based system was used to generate a biased rhizosphere [73]. It has been well documented for years that opines play a role in *Agrobacterium*–plant interactions. The researchers inoculated a transgenic plant producing an opine with *Pseudomonas* strains that either could or could not metabolize the opine. The opine-metabolizing strain was preferentially selected in the rhizosphere, achieving a population two to three times that of the non-opine-metabolizing strain.

In this manner, plants may be engineered to select for microbes of choice in the rhizopshere. Although the construction of engineered rhizosphere is focused with respect to agricultural applications, it may also be applied for remediation of contaminated soils using specific plant-microbe combinations.

Bacterial *mer A* gene was transferred to *Arabidopsis thaliana* [11] is another example of successful engineering of plants for phytoremediation. The protein is an oxidoreductase which catalyses the conversion of toxic Hg^{2+} to less toxic metallic mercury. The seeds of transformed plants were found to germinate and grow on medium containing up to 100 mM Hg, and evolved two to three times the amount of Hg^0 compared to control plants. Plants were also resistant to toxic levels of Au^{3+}. The group also reported the use of transgenic yellow poplar (*Liriodendron tulipifera*) for mercury phytoremediation and the transgenic plants were found to yield 10 times the amount of Hg^0 compared to control plants [12]. The system has not been tested under field conditions, but these studies indicate that genetic engineering may improve the capacity of plants to remediate metal-polluted soils.

Most of the bacteria possess ACC deaminase activity which promotes plant growth but it has been demonstrated that constitutive expression of ACC deaminase in transgenic plants offers some advantages over that in bacteria. The ACC deaminase activity in transgenic plants is higher compared to that in bacteria during initial stages of seed germination. Also it can constantly stimulate plant growth which leads to higher metal accumulation and in some cases an increase in the shoot/root ratio. Thus, it could improve the metal uptake of certain fast-growing plants for the substitution of slow-growing hyperaccumulators. Thus, development of transgenic plants carrying plant growth promoting property from rhizosphere bacteria holds great potential in plant development and phytoremediation.

Iron in soils is generally present as Fe (III) which is unavailable to plants. Different mechanisms for iron sequestration by plants include proton extrusion, siderophore production, and reduction of Fe (III) to Fe (II) by Fe (III) reductase. Another example of use of microbial genes for phytoremediation of metals is genetic engineering of tobacco plants carrying the Fe (III) reductase gene from *Saccharomyces cerevisiae*. The membrane-bound enzyme from the superfamily of flavochromes reduces Fe (III) to more soluble Fe (II). Thus, the transgenic plants were found to accumulate more iron and were tolerant to high Fe concentrations.

Genetic engineering of plants has been performed to carry MT genes from plant and animal origins which resulted in an increase in resistance to toxic metals. Various MT genes including mouse MTI, human MTIA (alpha domain), human MTII, Chinese hamster MTII, yeast CUP1, and pea PsMT have been transferred to *Nicotiana* sp., *Brassica* sp. and *A. thaliana* [74] resulting in enhanced tolerance to cadmium, maximum being 20-fold compared to the control. Transgenic plants carrying a MT gene of plant origin has also been reported. MT genes from Pea (*Pisum sativum*) PsMTA were expressed in *A. thaliana* and the transformed plants were reported to accumulate more Cu (several-fold in some plants) than the control plants [75].

Plants have also been engineered genetically to carry a rhamnolipid gene from bacteria. Researchers at the Sainsbury Laboratory (Norwich, UK), in collaboration with the Institute of Genetics and Cytology (Minsk, Belarus), have successfully produced rhamnolipids in transgenic plants expressing genes derived from the soil bacterium *Pseudomonas aeruginosa*. Transgenic *Arabidopsis* and tobacco plants were generated expressing individual or combinations of different *P. aeruginosa* genes (*e.g., rhl*A and *rhl*B) responsible for the biosynthesis of rhamnolipids. The transgenic plants producing rhamnolipids were found to exhibit dramatically enhanced growth and contamination tolerance on crude oil and heavy-metal contaminated soils, and to significantly enhance the biodegradation of crude oil. Increased resistance/tolerance was also exhibited to bacterial and fungal pathogens including *Phytophthora infestans* [76].

We have discussed some of the examples showing genetic engineering has been applied for the betterment of phytotechnologies for remediation of metals. However, the potential for using genetic resources in transgenic phytoremediation approaches has not yet been fully explored. Furthermore, a number of transgenic plants have been engineered to contain large amounts of recombinant proteins with a possible role in chelation, assimilation or membrane transport of trace elements. It is often difficult to predict the effects of the expression of a transgene at the level of the whole plant. An improved understanding of metal homeostasis in plants will be vital for the development of successful phytoremediation technologies.

CONCLUSIONS

Phytoremediation technologies hold great potential for restoration of land contaminated with metals. Remediation using phytotechnologies not only result in elimination of metals and/or their toxic effects but also helps to create a green zone. The presence of vegetation helps in improving soil quality and increases soil microbial diversity. Therefore, it is an environmentally friendly and aesthetic approach for remediation of metals as well as other pollutants. Rhizosphere microorganisms play an important role in phytoremediation technologies. Presence of beneficial microorganisms in the soils facilitates plant growth and thus increases phytoremediation efficiency. Their presence, regardless of the precise effects causes enhanced accumulation of metals in plants without decline in the biomass which is a fundamental need for effective phytoremediation. In this regard, heavy metals may be removed from polluted soils either by increasing the metal-accumulating ability of plants or by increasing the amount of plant biomass. In soils heavily contaminated with toxic metals, the metal contents exceed the limit of plant tolerance. It may be possible to treat plants with plant growth-promoting rhizobacteria to increase plant biomass and thereby stabilize, revegetate, and remediate metal-polluted soils.

In rhizosphere, the microrganisms and plants interact in such a manner that they create conditions which favour removal of toxic contaminants. These plant-microbe interactions can be enhanced or modulated by modifying microbial population (rhizoengineering) for the remediation of pollutants present in the soils. Rhizoengineering is an innovative approach towards phytoremediation.

In spite of many reports on the beneficial effects of soil microrganisms on phytoremediation, poor understanding in many relevant areas limits the successful exploitation of the technology. Complex microbe-plant and microbe-soil interactions in the rhizosphere need to be studied in more details. Similarly, there is a need to get insights on how these interactions affect the speciation of metals and thus their bioavailability.

ACKNOWLEDGEMENT

Declared none.

CONFLICT OF INTEREST

The author(s) confirm that this chapter content has no conflict of interest.

REFERENCES

[1]   Gade LH. Highly polar metal—metal bonds in "early-late" heterodimetallic complexes. Angewandte Chemie-International Edition 2000; 39:2658-78.

[2]   McKeehan P. Brown fields: The financial, legislative and social aspects of the redevelopment of contaminated commercial and industrial properties. 2000. Available from: http://www.csa.com/discoveryguides/brown/abstracts-f.php

[3]   Singh M. Heavy metal pollution in freshly deposited sediments of the Yamuna River (the Ganges River tributary): A case study from Delhi and Agra urban centres, India. Environ Geol 2001; 40: 664-71.

[4]   Kurek E, Bollag JM. Microbial immobilization of cadmium released from CdO in the soil. Biogeochemist 2004; 69:227-39.

[5]   Schwartz C, Echevarria G, Morel JL. Phytoextraction of cadmium with *Thlaspi caerulescens*. Plant Soil 2003; 249: 27-35.

[6]   Salt DE, Smith RD, Raskin I. Phytoremediation. Annu Rev Plant Physiol Plant Mol Biol 1998; 49: 643-68.

[7]   Raskin I, Kumar PBAN, Dushenkov S, Salt D. Bioconcentration of heavy metals by plants. Curr Opinion Biotechnol 1994; 5:285-90.

[8]   Environmental Protection Agency. Introduction to Phytoremediation (EPA-600-R-99-107). USA 2000; pp. 1-93.

[9]   Banuelos GS, Gardon G, Mackey B, *et al*. Boron and selenium removal in boron-laden soils by four sprinkler irrigated plant species. J Environm Qual 1993; 22:786-92.

[10]   De Souza MP, Chu D, Zhao M, *et al*. Rhizosphere bacteria enhance selenium accumulation and volatilization by Indian mustard. Plant Physiol 1999; 119:563-73.

[11]   Rugh CL, Wilde D, Stack NM, Thompson DM, Summers AO, Meagher RB. Mercuric ion reduction and resistance in transgenic *Arabidopsis thaliana* plants expressing a modified bacterial merA gene. PNAS USA 1996; 93: 3182-7.

[12]   Rugh CL, Senecoff JF, Meagher RB, Merkle SA. Mercury detoxification with transgenic plants and other biotechnological breakthroughs for phytoremediation. *In Vitro* Cellular & Develop Biol – Plant 1998; 37: 321-5.

[13]   ITRC (Interstate Technology & Regulatory Council). Phytotechnology technical and regulatory guidance and decision trees, Revised. Phyto-3 2009.

[14]   Liao SW, Chang WL. Heavy metal phytoremediation by water hyacinth at constructed wetlands in Taiwan. J Aquat Plant Manag 2004; 42: 60-8.

[15]   Francesconi K, Visootiviseth P, Sridokchan W, Goessler W. Arsenic species in an arsenic hyperaccumulating fern, *Pityrogramma calomelanos*: potential phytoremediator of arsenic-contaminatedsoil. Sci Total Environ 2002; 284:27-32.

[16]   Salt DE, Blaylock M, Kumar NP, Dushenkov V, Ensley BD, Chet I, Raskin I. Phytoremediation: a novel strategy for the removal of toxic metals from the environment using plants. Bio/Technol 1995; 13:468-74.

[17]   Dushenkov S, Vasudev D, Kapulnik Y, *et al*. Removal of uranium from water using terrestrial plants. Environ Sci Technol 1997; 31:3468–74.

[18]   Chandra P, Sinha S, Rai UN.Bioremediation of Chromium from Water and Soil by Vascular Aquatic Plants. In: Kruger EL, Anderson TA, Coats JR. Eds. Phytoremediation of Soil and Water Contaminants, Washington, ACS Symposium Series 664, 1997; p. 19.

[19]   Huang JW, Chen J, Berti WR, Cunningham SD. Phytoremediation of lead contaminated soils. Role of synthetic chelates in lead phytoextraction. Environ Sci Technol 1997; 31: 800-6.

[20]   Dushenkov V, Kumar PB, Motto AH, Raskin I. Rhizofiltration: the use of plant to remove heavy metals from aqueous streams. Environ Sci Technol 1995; 29:1239-45.

[21]   Peer WA, Baxter IR, Richards EL, Freeman JL, Murphy AS Phytoremediation and Hyperaccumulator Plants. In: Tamas M, Martinoia E, Eds. Molecular Biology of Metal Homeostasis and Detoxification. Topics in Current Genetics Vol 14, Berlin, Springer, 2005; pp. 299-340.

[22]   Murphy A, Zhou J, Goldsbrough PB, Taiz L. Purification and immunological identification of metallothioneins 1 and 2 from *Arabidopsis thaliana*. Plant Physiol 1997; 113: 1293–301.

[23]   Howden R, Cobbett CS. Cadmium-sensitive mutants of *Arabidopsis thaliana*. Plant Physiol 1992; 100:100-7.

[24]   Hall JL. Cellular mechanisms for heavy metal detoxification and tolerance. J Exp Bot 2002; 53:1-11.

[25] Lobreaux S, Briat JF Ferritin accumulation and degradation in different organs of pea (*Pisum sativum*) during development. Biochem. J 1991; 274: 601–6.

[26] Curl EA, Truelove B. The Rhizosphere. Springer-Verlag, Berlin, Germany 1986.

[27] Berg G, Eberl L, Hartmann A. The rhizosphere as a reservoir for opportunistic human pathogenic bacteria. Environ Microbiol 2005; 11:1683-5.

[28] Abou-Shanab RAI, Berkum P van, Angle JS. Heavy metal resistance and genotypic analysis of metal resistance genes in gram-positive and gram-negative bacteria present in Ni-rich serpentine soil and in the rhizosphere of *Alyssum murale*. Chemosphere 2007; 68: 360-7.

[29] Joshi PM, Juwarkar AA. *In vivo* studies to elucidate the role of extracellular polymeric substances from *Azotobacter* in immobilization of heavy metals. Environ Sci Technol 2009; 43:5584-9.

[30] Glick BR. The enhancement of plant growth by free-living bacteria. Can J Microbiol 1995; 41:109–17.

[31] Glick BR, Penrose DM, Li J P A model for the lowering of plant ethylene concentrations by plant growth-promoting bacteria. J Theor Biol 1998; 190: 63–8.

[32] Abou-Shanab RAI, Angle JS, Delorme TA, Chaney RL, Berkum P van, Moawad H, Ghanem K, Ghozlan HA. Rhizobacterial effects on nickel extraction from soil and uptake by *Alyssum murale*. New Phytol 2003; 158: 219-24.

[33] Smith S E, Read D J. Mycorrhizal Symbiosis. Academic Press, San Diego, California, USA 1997.

[34] Rajkumar M, Ae N, Prasad MN, Freitas H. Potential of siderophore-producing bacteria for improving heavy metal phytoextraction. Trends in Biotechnol 2010; 28: 142-9.

[35] Gamalero E, Trotta A, Massa N, Copetta A, Martinotti MG, Berta G. Impact of two fluorescent pseudomonads and an arbuscular mycorrhizal fungus on tomato plant growth, root architecture and P acquisition. Mycorrhiza 2004; 14:185-92.

[36] Oostendorp M, Sikofu RA. *In-vitru* interrelationships between rhizssphere bacteria and *Heterodera schachtii*. Revue Nematol 1990; 269-74.

[37] Glick BR, Patten CL, Holguin G, Penrose DM. Biochemical and Genetic Mechanisms Used by Plant Growth Promoting Bacteria. London, Imperial College Press, 1999; p. 270.

[38] Grichko VP, Glick BR. Amelioration of flooding stress by ACC deaminasecontaining plant growth-promoting bacteria. Plant Physiol Biochem 2001; 39:11-7.

[39] Grichko VP, Glick BR. Flooding tolerance of transgenic tomato plants expressing the bacterial enzyme ACC deaminase controlled by the 35*S*, *rolD* or PRB-1*b* promoter. Plant Physiol Biochem 2001b; 39:19-25.

[40] Penrose DM, Glick BR (2001) Levels of 1-aminocyclopropane-1-carboxylic acid (ACC) in exudates and extracts of canola seeds treated with plant growth-promoting bacteria. Can. J. Microbiol. 47:368-372.

[41] Evers A, Hancock RD, Martell AE, Motekaitis RJ. Metal ion recognition in ligands with negatively charged oxygen donor groups. Complexation of Fe(III), Ga(III), In(III), Al(III) and other highly charged metal ions. Inorg Chem 1989; 28:2189–95.

[42] Nair A, Juwarkar AA, Singh SK. Production and characterization of siderophores and its application in arsenic removal from contaminated Soil. Water Air Soil Pollut 2007; 180: 199-212.

[43] Rajkumar M, Freitas H. Effects of inoculation of plant-growth promoting bacteria on Ni uptake by Indian mustard. Bioresource Technol 2008; 99: 3491–8.

[44] Rajkumar M, Freitas H. Influence of metal resistant-plant growth promoting bacteria on the growth of *Ricinus communis* in soil contaminated with heavy metals. Chemosphere 2008; 71: 834–42.

[45] Barzanti R, Ozino F, Bazzicalupo M, *et al*. Isolation and characterization of endophytic bacteria from the nickel hyperaccumulator plant *Alyssum bertolonii*. 2007; 53:306-16.

[46] Idris R, Trifonova R, Puschenreiter M, Wenzel WW, Sessitsch A. Bacterial communities associated with flowering plants of the Ni hyperaccumulator Thlaspi goesingense. Appl Environ Microbiol 2004; 70: 2667-77.

[47] Rodriguez H, Gonzalez T, Goire I, Bashan Y. Gluconic acid production and phosphate solubilization by the plant growth-promoting bacterium *Azospirillum* spp. Naturwissenschaften 2004; 91:552–5.

[48] Kamnev AA. Reductive solubilization of Fe (III) by certain products of plant and microbial metabolism as a possible alternative to siderophore secretion. Doklady Biophys 1998; 358-360: 48-51.

[49] Leinhos V, Bergmann H. Influence of auxin producing rhizobacteria on root morphology and nutrient accumulation of crops. 2. Root-growth promotion and nutrient accumulation of maize (*Zea-mays* L.) by inoculation with indole-3-acetic acid (IAA) producing *Pseudomonas* strains and by exogenously applied IAA under different water-supply conditions. Angew Bot 1995; 69:37–41.

[50] Lippmann B, Leinhos V, Bergmann H Influence of auxin producing rhizobacteria on root morphology and nutrient accumulation of crops. 1. Changes in root morphology and nutrient accumulation in maize (*Zea-mays* L.) caused by inoculation with indole-3-acetic acid (IAA) producing *Pseudomonas* and *Acinetobacter* strains or IAA applied exogenously. Angew Bot 1995; 69:31–6.

[51] Zehnder GW, Murphy JF, Sikora EJ, Kloepper JW. Application of rhizobacteria for induced resistance. Euro J Plant Pathol 1997; 107:39-50.

[52] Nie L, Shah S, Burd GI, Dixon DG, Glick BR. Phytoremediation of arsenate contaminated soil by transgenic canola and the plant growth-promoting bacterium *Enterobacter cloacae* CAL2. Plant Physiol Biochem 2002; 40:355–61.

[53] Alstrom S. Induction of disease resistance in common bean susceptible to hal blight bacterial pathogen after seed bacterization with rhizosphere *Pseudomonas*. J Gen Appl Microbiol 1991; 37:495-501.

[54] Blake RC, Choate DM, Bardhan S, Revis N, Barton LL, Zocco TG. Chemical transformation of toxic metals by a *Pseudomonas* strain from a toxic waste site. Environ Toxic Chem 1993; 12:1365-76.

[55] Park CH, Keyhan M, Matin A Purification and characterization of chromate reductase in *Pseudomonas putida*. Abs Gen Meet Am Soc Microbiol 1999; 99:536-41.

[56] Delorme TA, Gagliardi JV, Angle JS, Chaney RL. Influence of the zinc hyperaccumulator *Thlaspi caerulescens* J. & C. Presl. and the nonmetal accumulator *Trifolium pratense* L. on soil microbial populations. Can J Microbiol 2001; 47:773–6.

[57] Mengoni A, Bazzicalupo M, Reeves RD, *et al*. Evolutionary dynamics of nickel hyperaccumulation in *Alyssum* revealed by its nrDNA analysis. New Phytol 2001; 159:691-9.

[58] Whiting SN, De Souza MP, Norman T. Rhizosphere bacteria mobilize Zn for hyperaccumulation by *Thlaspi caerulescens*, Environ Sci Technol 2001; 35:3144–50.

[59] Abou-Shanab B, Adwan G, Abu-Safiya D, Adwan G, Abu-Shanab M. Antibacterial activity of *Rhus coriaria*. L extracts growing in Palestine. J Islamic Univ Gaza (Natural Sciences series) 2005; 13: 147-53.

[60] Zhu Y, Geng C, Tong Y, Smith SE, Smith FA. Phosphate (Pi) and arsenate uptake by two wheat (*Triticum aestivum*) cultivars and their doubled haploid lines. Ann Bot 2006; 98:631-6.

[61] Leggett JE, Epstein E (1956) Kinetics of sulfate absorption by barley roots. Plant Physiol 1956; 31:222–6.

[62] Mason RP, Reinfelder JR, Morel FMM. Uptake, toxicity and trophic transfer of mercury in a coastal diatom. Environ. Sci Technol 1996; 30:1835–45.

[63] Lovley DR. Bioremediation of organic and metal contaminants with dissimilatory metal reduction. J Ind Microbiol 1995; 14:85–93.

[64] Wang YT, Shen H. Bacterial reduction of hexavalent chromium. J Ind Microbiol 1995; 14:159–63.

[65] Juwarkar AA, Nair A, Dubey KV, Singh SK, Devotta S. Biosurfactant technology for remediation of cadmium and lead contaminated soils. Chemosphere 2007; 68:1996-2002.

[66] Wang S, Mulligan CN. Rhamnolipid biosurfactant-enhanced soil flushing for the removal of arsenic and heavy metals from mine tailings. Process Biochem 2009; 44:296-301.

[67] Sheng X, He L, Wang Q, Ye H, Jiang C. Effects of inoculation of biosurfactant-producing *Bacillus* sp. J119 on plant growth and cadmium uptake in a cadmium-amended soil. J Hazard Mat 2008; 155:17-22.

[68] Garrett SD. Soil Fungi and Soil Fertility, 2nd Ed. Pergamon Press. Oxford, 1981; p. 77.

[69] Blaylock MJ, Huang JW. Phytoextraction of Metals. In: Raskin I, Ensley BD, Eds. Phytoremediation of Toxic Metals: Using Plants to Clean Up the Environment. New York, John Wiley & Sons Inc, 2000; pp. 193–229.

[70] Citterio S, Prato N, Fumagalli P, *et al*. The arbuscular mycorrhizal fungus *Glomus mosseae* induces growth and metal accumulation changes in *Cannabis sativa* L. Chemosphere 2005; 59:21–9.

[71] Pierson EA, Wood DW, Cannon JA, Blachere FM, Pierson LS III. Interpopulation signaling *via* N-acyl-homoserine lactones among bacteria in the wheat rhizosphere. Mol Plant Microbe Interact 1998; 11:1078–84.

[72] Fray RG, Throup JP, Daykin M, *et al*. Plants genetically modified to produce *N*-acyl homoserine lactones communicate with bacteria. Nature Biotechnol 1999; 17:1017-20.

[73] Savka MA, Farrand SK. Modification of rhizobacterial populations by engineering bacterium utilization of novel plant-produced resource. Nat Biotechnol 1997; 15: 363–8.

[74] Karenlampi S, Schat H, Vangronsveld J, *et al*. Genetic engineering in the improvement of plants for phytoremediation of metal polluted soils. Environ Pollut 2000; 107:225– 31.

[75] Evans KM, Gatehouse JA, Lindsay WP, Shi J, Tommey AM, Robinson NJ. Expression of the pea metallothionein like gene Ps MTA in *Escherichia coli* and *Arabidopsis thaliana* and analysis of trace metal ion accumulation:implications of Ps MTA function. Plant Mol Biol 1992; 20:1019– 28.

[76]   Brychkova I. Bioremediation of oil and metal co-contaminated soil by plants producing rhamnolipid biosurfactants. 7th Biennial Symposium of Int Soc Environ Biotech, Chicago: USA 2004; pp. 18-21.

CHAPTER 2

Chelate-Assisted Phytoremediation of Lead

Lingjuan Zheng[1,*], Ursula Lütz-Meindl[2], Thomas Peer[1]

[1]Department of Organismic Biology and [2]Department of Cell Biology, University of Salzburg, Hellbrunnerstraße 34, 5020 Salzburg, Austria

Abstract: Success of phytoremediation greatly depends on the plant species. In addition, bioavailability of Pb in soils is regarded as the key factor limiting the efficiency of phytoextraction. Different biological, physical and chemical methods including addition of chelating agents such as EDTA and other bio-degradable chelators have been investigated. In the absence of added chelating agents in soil, only few species are able to achieve the status of a Pb hyperaccumulator. In contrast, some studies showed that plants exposed to Pb with supplement of EDTA were able to uptake from dozen to hundred times more Pb in the shoots than those treated with Pb alone. Addition of chelating agents combined with application of electric current or plant growth regulators, might greatly increase bioavailability of Pb in soils and / or plant biomass, thus ultimately the efficiency of phytoextraction. In plants treated with Pb plus EDTA, Pb deposits at the ultrastructural level were found mainly in cell walls, vacuoles and along plasma membranes in various patterns (acicular, granular and fine precipitates) in root cells of *Lespedeza chinensis* and *L. davidii*. These might be related to the transport and detoxificaiton of Pb chelates in the plants.

Keywords: Heavy metal, lead phytoremediation, phytoextraction, chelating agents, ethylenediamine tetraacetic acid (EDTA), ethylenediamine disuccinic acid (EDDS), N-(2-hydroxyethyl) ethylenediaminetriacetic acid (HEDTA), bio-degradable chelators, plant hormones, indole-3-acetic acid (IAA), microbial inoculation, arbuscular mycorrhizal fungi, detoxification mechanism, ultrastructure, localization, transmission electron microscopy - electron energy loss spectroscopy (TEM), cell wall, vacuole, apoplastic and symplastic pathways, X-ray diffraction (XRD).

INTRODUCTION

Since the industrial revolution, the production of heavy metals has increased

---

*Address correspondence to Lingjuan Zheng: Department of Organismic Biology, University of Salzburg, Hellbrunnerstraße 34, 5020 Salzburg, Austria; Tel: 0043 662 8044 5529; Fax: 0043 662 8044 142; E-mail: lithops2009@hotmail.com

exponentially, with serious consequences for terrestrial and aquatic ecosystems, as well as human health [1-3]. Approximately more than 16% of the total land area (estimated 52 million hectares) is affected by soil contamination [4], and the worldwide release of Pb reached up to 783,000 t over the past five decades [5]. According to the CERCLA Priority List of Hazardous Substances by Agency for Toxic Substances and Disease Registry (ATSDR), Pb is regarded as one of the most environmentally hazardous substances because of the prevalence and severity of the toxicity (http://www.atsdr.cdc.gov/cercla/07list.html). Anthropogenic pollution of soil with Pb not only results in the sharp decrease of agricultural production in terms of quantity and quality, but also poses a severe threat to human health by the biomagnifications through food chain [6, 7]. Therefore, pollution control and remediation of Pb are urgent.

In comparison with conventional remediation methods like washing, solvent extraction, thermal desorption, and flotation [8, 9], phytoremediation is a much more environmentally friendly and cost-effective way to decontaminate soil. Phytoremediation of heavy metals, such as Hg, Cd, Cu, Ni, Pb, Cr, Mn and Zn, has been widely studied in last several decades (*e.g.,* [10-12]). Searching for "phytoremediation, chelating agents, heavy metal and Pb" in Google Scholar, has found that about 4,500 citations are actually available and more findings are forthcoming. There have been many reviews written about the topic but it seems impossible to give a single comprehensive overview on this topic. Here, a brief introduction and an update on chelate-assisted phytoremediation of Pb from our own research will be provided.

Phytoremediation is a general term including five processes: phytoextraction, phytofiltration, phytostabilization, phytovolatilization and phytodegradation [13], among which phytostabilization and phytoextration are the most widely studied approaches for Pb-contaminated soils [14, 15]. Phytostabilization involves the use of plant roots to absorb the pollutants from soils and keep them as immobilized and harmless forms in the rhizosphere. In this way, the pollutants can be prevented from leaching [16]. Since phosphate compounds of Pb are comparatively insoluble, it has been suggested that inducing the formation of phosphate compounds may offer a solution to reduce the bioavailability of Pb in soil [17]. In a latter study using analytical transmission electron microscopy

(ATEM) and X-ray absorption spectroscopy (XAS), it was proven that apatite can form pyromorphite in soil [18].

Phytoextraction has often been proposed as a promising technique to remediate lands contaminated with heavy metals [19, 20], first involving direct uptake and accumulation of heavy metals in plant tissues with subsequent removal of the plants [5]. There are two strategies for phytoextraction: continuous and induced phytoextraction [21]. Continuous phytoextraction requires use of wild hyperaccumulator plants to accumulate heavy metals in harvestable plant tissues. Induced phytoextraction requires the addition of other substances such as chelating agents, microbes and plant hormones. These additives are intended to mobilize heavy metals or enhance biomass of plants so that uptake and translocation of heavy metals by plants can be accelerated [9, 22, 23]. In the following, a short summary focusing on chelate-assisted Pb phytoremediation and the associted detoxification mechanisms at subcellular level will be given.

CHELATE-ASSISTED PHYTOREMEDIATION OF LEAD

Success of phytoextraction greatly depends on the plant species. A number of greenhouse studies and field surveys have been carried out on this issue (*e.g.,* [11, 24, 25]). Species from various families have been investigated for Pb phytoextraction, such as *Brassica juncea, Bidens maximowicziana, Lespedeza davidii, Medicago sativa, Phaseolus vulgaris, Sesbania drummondii, Vetiveria zizanoides*, and *Zea mays* [12, 25-31]. However, in the absence of chelating agents in soil, only few species are able to achieve the definition as a Pb hyperaccumulator (*e.g.,* [25, 26, 29]). In Table **1**, there is a summary of some plant species which have been proven to be able to accumulate higher than 1000 ppm of Pb in their shoots under laboratory conditions, with the addition of substances such as chelating agents, plant hormones, *etc.* In all species investigated, *Brassica juncea, Zea mays, Phaseolus vulgaris*, and *Typha orientalis* showed much higher Pb tolerance and accumulation ability than others.

Apart from the plant species used for phytoextraction, bioavailability of Pb in soils is regarded as the key factor limiting the efficiency of phytoextraction. Generally, the bioavailability of Pb in soils is very low and depends on physiochemical properties of soil such as pH, clay content and organic matter content [32, 33]. According to the sequential extraction procedures, heavy metals in soils are divided into five important fractionations: exchangeable, carbonate bound, Fe- and Mn-oxides bound, organic bound, and residual portions [34], in which only the exchangeable and carbonate bound fractions are regarded to be available for plant uptake. The forms of Pb available for plant uptake vary from 0% to 1.68% [35]. Therefore, efforts on increasing Pb bioavailability in soil by applying different physical, chemical, biological, and combined methods were made, in order to improve the efficiency of phytoextraction [9]. Supplement of proper electric current in soil can stimulate the accumulation of multiple metals in Indian mustard (*Brassica juncea*) including Pb [36]. Combined use of electric potential difference application and EDTA on *Brassica juncea* induced a 4-fold increase in Pb uptake and a significant reduction in leaching, compared to control [37].

The potential of various chelating agents for Pb phytoremediation such as EDTA, EDDS, HEDTA, nitrilotriacetic acid (NTA), diethylene trinitrilopentaacetic acid (DTPA) and citric acid (CA), have been widely investigated in last decades [29, 38]. A study on potential of *Hemidesmus indicus* for Pb phytoextraction with addition of EDTA, HEDTA, DTPA and *trans*-1,2-diaminocyclohexane-*N,N,N′,N′*-tetraacetic acid (CDTA) showed that application of HEDTA and EDTA increased removal of Pb from contaminated soil significantly, whereas the supplement of DTPA and CDTA improved Pb contents in roots but not in shoots, thus hindered Pb accumulation in whole plant in total [39].

In most soil amendments studied, EDTA is often proven to be the most efficient chelating agent for Pb phytoremediation [26]. In comparison with plants under Pb treatment alone, plants with supplement of EDTA are able to uptake from dozen to hundred times more Pb in the shoots (Table **1**). EDTA solubilizes Pb mainly from carbonate-specifically adsorbed and Fe- and Mn-oxide complexes in soils [35, 40]. However, the rapid increase of soluble Pb by addition of EDTA in soils may pose a potential risk of groundwater pollution by leaching. Xie *et al.* [41] used a controlled-release microencapsulated EDTA which can provide lower and more constant water-soluble Pb to remediate Pb-contaminated soil with maize. Moreover, the non-biodegradability of EDTA inhibits its field application for Pb

Table 1: Summary of plant species that are suggested to be used for chelate-assisted Pb phytoextraction

Species	Concentration of applied chelating agents	Chelating agents	Highest Pb concentra-tion in shoots (mg.kg^{-1})	Times of Pb accumulation in shoots by addition of chelating agents	Culture	References
Bidens maximowicziana	3.15 mmol.kg^{-1}	EDTA	1906	2.8	pot experiment	[30]
Brassica juncea	0.75 mM	EDTA	11592	400	hydroponic	[27]
	10 mmol.kg^{-1}		1923	116	pot experiment	[59]
	0.5 mmol.kg^{-1}		5000	22.5	pot experiment	[65]
Cynara cardunculus	1 g.kg^{-1}	EDTA	1332	30.8	pot experiment	[66]
Zea mays L. cv. Fiesta	0.5 g.kg^{-1}	HEDTA	1110 μg.plant^{-1}	26.4	pot experiment	[26]
		EDTA	2410 μg.plant^{-1}	57.4	pot experiment	
Zea mays L. cv. Fiesta	2 g.kg^{-1}	HEDTA	10600	265	pot experiment	[67]
Zea mays cultivar AG1051	20 mmol.kg^{-1}	EDTA	~2400	~81	pot experiment	[68]
Brassica rapa L. subsp. chinensis (L.) Hanelt cv. Xinza No 1	3 mmol.kg^{-1}	EDTA	5010	39.8	pot experiment	[40]
Vigna radiata (L.) R. Wilczek var. radiata cv. VC-3762	3 mmol.kg^{-1}	EDTA	1170	9.2	pot experiment	
Triticum aestivum L. cv. Altas 66	3 mmol.kg^{-1}	EDTA	2650	33	pot experiment	
Triticum aestivum L.	3 mmol.kg^{-1}	EDTA	1095	60.8	pot experiment	[69]
Brassica juncea (L.) Czem.	8 mmol.kg^{-1} + 160 mmol.kg^{-1}	EDTA+S	7100	23.3	pot experiment	

Table 1: contd….

Plant species	Dose	Chelate	Pb concentration	Value	Experiment	Ref.
Festuca arundinacea Schreb.	5 mmol.kg⁻¹+ 5 mmol.l⁻¹	EDTA+Acetic acid	~5500	---	pot experiment	[70]
Hemidesmus indicus	1g.kg⁻¹	EDTA	5196	1.99	pot experiment	[39]
		HEDTA	4719	1.8		
Lespedeza chinensis	10 mmol.kg⁻¹	EDTA	2327 in leaves	32	pot experiment	[12]
Lespedeza davidii			5663 in leaves	47		
Medicago sativa	0.2 mM+100 µM	EDTA+IAA	~ 2600 in leaves	28	hydroponic	[28]
Medicago sativa	0.2 mM+100 µM	EDTA+KN	~3800 ppm in leaves	---	hydroponic	[71]
Mirabilis jalapa	8 mmol.kg⁻¹	EDDS	5700 in leaves	7.8	pot experiment	[72]
		MGDA	5500 in leaves	7.5		
Phaseolus vulgaris	1 mM	EDTA	9176	6.1	hydroponic	[25]
Pisum sativum L. cv. Sparkle	0.5 g.kg⁻¹	DTPA	1930 µg.plant⁻¹	23.8	pot experiment	[26]
		EGTA	2080 µg.plant⁻¹	25.7		
		HEDTA	5670 µg.plant⁻¹	70		
		EDTA	8960 µg.plant⁻¹	110.6		
Pelargonium zonale	5 mmol.kg⁻¹	EDTA	2291	3.4	pot experiment	[73]
Sesbania drummondii	10 mmol.kg⁻¹	EDTA	4000-5000	55	pot experiment	[29]
	5 mmol.kg⁻¹	DTPA	1000-2000	22.5		
	10 mmol.kg⁻¹	HEDTA	3000-4000	42.5		
	10 mmol.kg⁻¹	NTA	1000-2000	21.3		
	2.5 mmol.kg⁻¹	Citric acid	1000-2000	17.5		
Sesbania drummondii	100 µM+100 mg.l⁻¹	EDTA+IAA	2037	13.5	hydroponic	[23]
Sedum alfredii Hance	10 mmol.kg⁻¹	EDTA	1190	5.9	pot experiment	[74]

Table 1: contd....

Typha latifolia	15 mmol.kg⁻¹	EDTA	2418	---	pot experiment	[75]
Typha orientalis Presl	0.1 mM	EDTA	~21,500 in leaves	1.3	hydroponic	[76]
Zinnia elegans Jacq.	4.8 mM	EDTA	6752 in seedling	1.4	hydroponic	[77]
	4.8 mM	Tartaric acid	6541 in seedling	1.3	hydroponic	
	2.4 mM	Citric acid	6598 in seedling	1.3	hydroponic	
Fagopyrum esculentum Moench. cv. Pingqiao No. 2	2.5 mmol.kg⁻¹	EDTA	2500	---	pot experiment	[78]
Pisum sativum L. cv. Qinxuan No. 2			1110			
Helianthus annuus L. cv. S61	5.0 mmol.kg⁻¹		1800			
Brassica juncea L. Czern. et Coss. cv. Liyangkucai			2900			

Abbreviations: EGTA - Ethylene glycol tetraacetic acid; KN - Kinetin; MGDA - Methylglycinediacetic acid; S - Sulfur.

phytoremediation. Nevertheless, completely opposite effect of EDTA on Pb phytoremediation was also reported. Tian *et al.* [42] found that the intensities of Pb were lower in *Sedum alfredii* treated with EDTA and Doncheva *et al.* [43] showed lower Pb contents in two sunflower genotypes with addition of EDTA. Such controversial results could be caused by plant species, Pb and chelators' concentrations applied and experimental conditions, *etc.*

Recently, there has been more research done on bio-degradable chelating agents. Elless *et al.* [14] found that EDDS and dicarboxymethyl glutamic acid tetrasodium salt (GLDA) are able to improve Pb solubility than EDTA in most soils from various urban locations in the United States, depending on soil properties. Attention has also been extended to enhance plant biomass or Pb bioavailability in soils by supplements of plant hormones or microbial community in combination with chelating agents, and thus ultimately help improve the efficiency of Pb phytoremediation. In the study of Hadi *et al.* [44] on maize, the interactions of gibberellic acid (GA$_3$), IAA and EDTA in improving plant growth and phytoextraction of Pb was investigated. The EDTA treatment alone increased Pb uptake but inhibited plant growth. Foliar spray of GA$_3$ and IAA compensated for the negative effect of EDTA on plant growth and the highest Pb uptake was found under EDTA+GA$_3$ treatment. Microbial inoculation of Pb-contaminated soil can promote not only plant growth but also Pb accumulation by plant tissues. This could involve increased production of IAA, siderophores, and 1-aminocyclopropane-1-carboxylate (ACC) deaminase, or improvement in soil Pb bioavailability [45]. Moreover, the synergistic combination of microorganisms and chelators can enhance phytoremediation efficiency. In a study by Gao *et al.* [46], it was shown that 10-15% more Pb was acquired in *Solanum nigrum* in the presence of citric acid and metal-resistant microorganisms than in control plants.

DETOXIFICATION MECHANISMS OF CHELATED-LEAD IN PLANTS

Chelation and compartmentalization are regarded as two important detoxification mechanisms of heavy metals in plants [47]. Phytochelatins (PCs) are a set of heavy-metal-complexing peptides with the general structure of (γ-Glu-Cys)$_n$ Gly (n = 2 - 11) [48] whose synthesis can be induced by Pb, bind Pb ions and then be sequestrated in plants [49]. Nevertheless, Gupta *et al.* [50] pointed out that the detoxification of Pb is not related to PCs but to glutathione (GSH) in *Sedum alfredii*. In the aquatic fern *Salvinia minima*, after Pb treatment higher GSH content was also observed in leaves than in roots [51]. Interestingly, a previous study showed higher PCs synthesis in roots than in leaves [52], indicating that possible differential detoxification strategies existed, even in the different parts of a plant. In the absence of added chelating agents in soils, most Pb is generally restricted to plant roots because of the sequestration of Pb as insoluble forms into organelles and cell walls. In above-ground parts, Pb is mainly restricted in the vascular bundles and epidermis tissues [53, 54]. Pb was reported to be accumulated in the forms of lead-acetate and lead-sulfur in *Sesbania drummondii*, cerussite (lead carbonate) in *Phaseolus vulgaris*, chloropyromorphite [$Pb_5(PO_4)_3Cl$], possibly phosphohedyphane [$Pb_3Ca_2(PO_4)_3Cl$] in *Brassica juncea* [55-57], and as a mixture of $Pb_3(PO_4)_2$, Pb-malic and Pb-GSH in *Sedum alfredii* [42]. In the presence of added chelating agents in soils, the accumulation and translocation of Pb can be facilitated markedly (*e.g.,* [27, 30, 38]). In this case, Pb remained soluble and was easily transported to above-ground parts of the plants. It was concluded that Pb was absorbed and translocated as Pb-EDTA complex following an analysis of xylem exudates [27, 59]. It was also proved by extended X–ray absorption fine structure spectroscopy (EXAFS) that Pb was predominantly accumulated as a mixture of Pb-EDTA in leaves of *Phaseolus vulgaris* and another species when grown in Pb-EDTA solution [55]. Although chelation and compartmentalization are the main detoxification mechanisms of Pb in plants, various Pb compounds identified in different plant species imply that plants may have multiple distinct strategies to avoid toxicity induced by Pb.

As added chelating agents are to increase Pb bioavailability, the formation of Pb precipitates in plant cells in the presence of added chelating agents must differ from the formation of Pb deposits in the absence of any added chelating agents. Transport of H-EDTA and EDTA-chelated Pb in *Chamaecytisus palmensis* and *Pinus radiata* was studied based on ultrastructural observations using transmission electron microscopy [58, 60]. These investigations showed differences in pattern and subcellular distribution of Pb precipitates which were regarded to be Pb in the root cells of these plants. In our previous study, we found that Pb was accumulated mainly in cell walls, vacuoles and along plasma membranes in various patterns (acicular, granular and fine precipitates) in root cells of *Lespedeza chinensis* and *L. davidii* under Pb plus EDTA treatment, suggesting both apoplastic and symplastic pathways of Pb transport in plants. However, Pb was observed in cell walls, vacuoles, intercellular space and middle lamella in the root cells of both species under Pb treatment alone. Moreover, differences in subcellular distribution of Pb were also observed between two species. By means of electron energy loss spectroscopy (EELS) analysis, Pb was identified to be sequestered primarily as lead phosphate in *L. chinensis* and *L. davidii* [61]. All of these data indicate differing tolerance and detoxification mechanisms depending on the use of chelating agents and plant species.

CONCLUSIONS

The ability for high uptake, accumulation and tolerance of metals by hyperaccumulating plants is an important consideration for Pb phytoremediation. In addition, the biomass of plants also plays a crucial role in storing and removing Pb or other toxic metals from soils. Therefore, the purposes of combined application of chemical, physical and biological methods (*e.g.,* usage of chelators, phytohormones, electric current, metal-resistance bacteria) for phytoremediation at present are to: 1) stimulate plant growth, and 2) improve Pb bioavailability in soil. Additionally, the potential of trees like poplar and willow for phytoremediation of metals has received increased attention in the past few decades. For techniques used for phytoremediation, recent applications of various techniques, detoxification mechanisms of plants, advantages and disadvantages of phytoremediation, see references [20, 45, 62-64].

With the development of molecular tools, the physiology of hyperaccumulators and the molecular mechanisms involved in metal resistance / tolerance as well as hyperaccumulation of metals has been discovered step by step. Transgenic plants expressed by the transformation of genes from other organisms or other plant species, or overexpression of genes from the same plant species, have been successfully applied to promote phytoremediation of metals from contaminated soils, but none have reached commercial use. In our study on *L. chinensis* and *L. davidii*, we found that cell wall sequestration and vacuolar compartmentalization are the two main mechanisms for Pb accumulation and detoxification in plants.

Therefore, further studies on identification of enzymes and genes contributing to cell wall sequestration, vacuolar compartmentalization, in particular vacuolar transporters and translocation of Pb from roots to leaves would have a great potential to assist the development of transgenic plants with superior remediation capacity. Until now, combined application of various methods mentioned above is mainly tested on native species from metal-polluted areas. It would be interesting to see if transgenic plants cannot only produce high biomass but can also accumulate and are tolerant to high concentrations of Pb in above-ground parts of the plants, in the presence of added chelators or plant growth regulators.

ACKNOWLEDGEMENTS

The authors wish to acknowledge the whole working groups of Prof. Thomas Peer and Prof. Ursula Lütz-Meindl for all support.

CONFLICT OF INTEREST

The author(s) confirm that this chapter content has no conflict of interest.

REFERENCES

[1]   Nriagu JO. A history of global metal pollution. Sci 1996; 272: 223-24.

[2]   Wei BG, Yang LS. A review of heavy metal contaminations in urban soils, urban road dusts and agricultural soils from China. Microchem J 2010; 94: 99-107.

[3]   Yabe J, Ishizuka M, Umemura T. Current levels of heavy metal pollution in Africa. J Veterin Med Sci 2010; 72: 1257-63.

[4]   Memon A, Schroeder P. Implications of metal accumulation mechanisms to phytoremediation. Environ Sci Pollut Res 2009; 16: 162-75.

[5]   Singh OV, Labana S, Pandey G, Budhiraja R, Jain RK. Phytoremediation: an overview of metallic ion decontamination from soil. Appl Microbiol Biotechnol 2003; 61: 405-12.

[6]   Laskowski R, Hopkin SP. Accumulation of Zn, Cu, Pb and Cd in the garden snail (*Helix aspersa*): Implications for predators. Environ Pollut 1996; 91: 289-97.

[7]   Zhao SP, Ye XZ, Zheng JC. Lead-induced changes in plant morphology, cell ultrastructure, growth and yields of tomato. Afri J Biotechnol 2011; 10: 10116-24.

[8]   Peng JF, Song YH, Yuan P, Cui XY, Qiu GL. The remediation of heavy metals contaminated sediment. J Hazard Mat 2009; 161: 633-40.

[9]   Karami A, Shamsuddin ZH. Phytoremediation of heavy metals with several efficiency enhancer methods. A J Biotechnol 2010; 9: 3689-98.

[10] Chaney RL, Malik M, Li YM, *et al*. Phytoremediation of soil metals. Curr Opin Biotech 1997; 8: 279-84.

[11]   Chandra R, Yadav S. Phytoremediation of Cd, Cr, Cu, Mn, Fe, Ni, Pb and Zn from aqueous solution using *Phragmites communis*, *Typha angustifolia* and *Cyperus esculentus*. International J Phytoremediat 2011; 13: 580-91.

[12]   Zheng LJ, Liu XM, Luetz-Meindl U, Peer T. Effects of lead and EDTA-assisted lead on biomass, lead uptake and mineral nutrients in *Lespedeza chinensis* and *Lespedeza davidii*. Water Air and Soil Pollut 2011; 220: 57-68.

[13]   Garbisu C, Alkorta I. Phytoextraction: a cost-effective plant-based technology for the removal of metals from the environment. Bioresour Technol 2001; 77: 229-36.

[14]   Elless MP, Bray CA, Blaylock MJ. Chemical behaviour of residential lead in urban yards in the United State. Environ Pollut 2007; 148: 291-300.

[15]   Lopareva-Pohu A, Verdin A, Garcon G, *et al.* Influence of fly ash aided phytostabilisation of Pb, Cd and Zn highly contaminated soils on *Lolium perenne* and *Trifolium repens* metal transfer and physiological stress. Environ Pollut 2011; 159: 1721-9.

[16]   Lone MI, He ZL, Stoffella PJ, Yang XE. Phytoremediation of heavy metal polluted soils and water: progresses and perspectives. J Zhejiang University-Sci B 2008; 9: 210-20.

[17]   Laperche V, Logan TJ, Gaddam P, Traina SJ. Effect of apatite amendments on plant uptake of lead from contaminated soil. Environ Sci Technol 1997; 31: 2745-53.

[18]   Cotter-Howells JD, Champness PE, Charnock JM. Mineralogy of Pb-P grains in the roots of *Agrostis capillaris* L. by ATEM and EXAFS. Mineralogical Magazine 1999; 63: 777-777(1).

[19]   Baker AJM., Brooks RR. Terrestrial higher plants which hyperaccumulate metal elements – A review of their distribution, ecology, and phytochemistry. Biorecovery 1989; 1: 81-126.

[20]   Evangelou MWH, Ebel M, Schaeffer A. Chelate assisted phytoextraction of heavy metals from soil. Effect, mechanism, toxicity, and fate of chelating agents. Chemosphere 2007; 68: 989-1003.

[21]   Salt DE, Smith RD, Raskin I. Phytoremediation. Annu Rev Plant Phys 1998; 49: 643-68.

[22]   Chen X, Wu CH, Tang JJ, Hu SJ. Arbuscular mycorrhizae enhance metal lead uptake and growth of host plants under a sand culture experiment. Chemosphere 2005; 60: 665-71.

[23]   Israr M, Sahi SV. Promising role of plant hormones in translocation of lead in *Sesbania drummondii* shoots. Environ Pollut 2008; 153: 29-36.

[24]   Rotkittikhun P, Kruatrachue M, Chaiyarat R, *et al.* Uptake and accumulation of lead by plants from the Bo Ngam lead mine area in Thailand. Environ Pollut 2006; 144: 681-8.

[25]   Piechalak A, Malecka A, Baralkiewicz D, Tomaszewska B. Lead uptake, toxicity and accumulation in *Phaseolus vulgaris* plants. Biol Plant 2008; 52: 565-8.

[26]   Huang JWW, Chen JJ, Berti WR, Cunningham SD. Phytoremediation of lead-contaminated soils: Role of synthetic chelates in lead phytoextraction. Environ Sci Technol 1997; 31: 800-5.

[27]   Vassil AD, Kapulnik Y, Raskin I, Salt DE. The role of EDTA in lead transport and accumulation by Indian mustard. Plant Physiol 1998; 117: 447-53.

[28]   López ML, Peralta-Videa JR, Benitez T, Gardea-Torresdey JL. Enhancement of lead uptake by alfalfa (*Medicago sativa*) using EDTA and a plant growth promotor. Chemosphere 2005; 61: 595-8.

[29]   Ruley AT, Sharma NC, Sahi SV, Singh SR, Sajwan KS. Effects of lead and chelators on growth, photosynthetic activity and Pb uptake in *Sesbania drummondii* grown in soil. Environ Pollut 2006; 144: 11-8.

[30]   Wang HQ, Lu SJ, Li H, Yao ZH. EDTA-enhanced phytoremediation of lead contaminated soil by *Bidens maximowicziana*. J Environ Sci-China 2007; 19: 1496-9.

[31]   Danh LT, Truong P, Mammucari R, Tran T, Foster N. Vetiver grass, *Vetiveria zizanioides*: a choice plant for phytoremediation of heavy metals and organic wastes. Int J Phytoremediat 2009; 11: 664-91.

[32]   Papafilippaki A, Gasparatos D, Haidouti C, Stavroulakis G. Total and bioavailable forms of Cu, Zn, Pb and Cr in agricultural soils: A study from the hydrological basin of Keritis, Chania, Greece. Global Nest J 2007; 9: 201-6.

[33]   Magrisso S, Belkin S, Erel Y. Lead bioavailability in soil and soil components. Water Air and Soil Pollut 2009; 202: 315-23.

[34]   Konradi EA, Frentiu T, Ponta M, Cordos E. Use of sequential extraction to assess metal fractionation in soils from Bozanta Mare, Romania. Acta Universitatis Cibiniensis Seria F Chemia 2005; 8: 5-12.

[35]   Leštan D, Grčman H, Zupan M, Bačac N. Relationship of soil properties to fractionation of Pb and Zn in soil and their uptake into *Plantago lanceolata*. Soil Sediment Contaminat 2003; 12: 507-22.

[36]   Cang L, Wang QY, Zhou DM, Xu H. Effects of electrokinetic-assisted phytoremediation of a multiple-metal contaminated soil on soil metal bioavailability and uptake by Indian mustard. Sep Purif Technol 2011; 79: 246-53.

[37]   Falciglia PP, Vagliasindi FGA. Enhanced phytoextraction of lead by Indian mustard using electric field. Chem Eng Trans 2013; 32: 379-84.

[38]   Tandy S, Bossart K, Mueller R, *et al.* Extraction of heavy metals from soils using biodegradable chelating agents. Environ Sci Technol 2004; 38: 937-44.

[39]   Sekhar KC, Kamala CT, Chary NS, Balaram V, Garcia G. Potential of *Hemidesmus indicus* for phytoextraction of lead from industrially contaminated soils. Chemosphere 2005; 58: 507-14.

[40]   Shen ZG, Li XD, Wang CC, Chen HM, Chua H. Lead phytoextraction from contaminated soil with high-biomass plant species. J Environ Qual 2002; 31: 1893-900.

[41]   Xie ZY, Wu LH, Chen NC, *et al.* Phytoextraction of Pb and Cu contaminated soil with maize and microencapsulated EDTA. Int. J. Phytoremediat 2012; 14: 727-40.

[42]   Tian SK, Lu LL, Yang XE, *et al.* The impact of EDTA on lead distribution and speciation in the accumulator *Sedum alfredii* by synchrotron X-ray investigation. Environ Pollut 2011; 159: 782-8.

[43]   Doncheva S, Moustakas M, Ananieva K, *et al.* Plant response to lead in the presence or absence EDTA in two sunflower genotypes (cultivated *H. annuus* cv. 1114 and interspecific line *H. annuus×H. argophyllus*). Environ Sci Pollut Res Int 2013; 20: 823-33.

[44]   Hadi F, Bano A, Fuller MP. The improved phytoextraction of lead (Pb) and the growth of maize (*Zea mays* L.): the role of plant growth regulators (GA_3 and IAA) and EDTA alone and in combinations. Chemosphere 2010; 80: 457-62.

[45]   Guptal D.K., Huang H.G., Corpas F.J. Lead tolerance in plants: strategies for phytoremediation. Environ Sci Pollut Res 2013; 20: 2150-61.

[46]   Gao Y., Miao C.Y., Wang Y.F. Xia J., Zhou P. Metal-resistant microorganisms and metal chelators synergistically enhance the phytoremediation efficiency of in Cd- and Pb-contaminated Soil. Environ Technol 2012; 33: 1383-9.

[47]   Yadav SK. Heavy metals toxicity in plants: An overview on the role of glutathione and phytochelatins in heavy metal stress tolerance of plants. S Afr J Bot 2010; 76: 167-79.

[48]   Zenk MH. Heavy metal detoxification in higher plants – a review. Gene 1996; 179: 21-30.

[49]   Sharma P, Dubey RS. Lead toxicity in plants. Braz J Plant Physiol 2005; 17: 35-52.

[50]   Gupta DK, Huang HG, Yang XE, Razafindrabe BHN, Inouhe M. The detoxification of lead in *Sedum alfredii* H. is not related to phytochelatins but the glutathione. J Hazard Mater 2010; 177: 437-44.

[51]   Estrella-Gómeza NE, Sauri-Duchb E, Zapata-Pérezc O, Santamaría JM. Glutathione plays a role in protecting leaves of *Salvinia minima* from Pb^{2+} damage associated with changes in the expression of SmGS genes and increased activity of GS. Environ. Exp Bot 2012; 75: 188-94.

[52]   Estrella-Gómeza N, Mendoza-Cózatlb D, Moreno-Sánchezb R, *et al.* The Pb-hyperaccumulator aquatic fern *Salvinia minima* Baker, responds to Pb^{2+} by increasing phytochelatins *via* changes in SmPCS expression and in phytochelatin synthase activity. Aquat Toxicol 2009; 91: 320-8.

[53]   Tian SK, Lu LL, Yang XE, Webb SM, Du YH, Brown PH. Spatial imaging and speciation of lead in the accumulator plant *Sedum alfredii* by microscopically focused synchrotron X-ray investigation. Environ Sci Technol 2010; 44: 5920-6.

[54]   Zhang J, Tian SK, Lu LL, Shohag MJI, Liao HB, Yang XE. Lead tolerance and cellular distribution in *Elsholtzia splendens* using synchrotron radiation micro-X-ray fluorescence. J Hazard Mater 2011; 197: 264-71.

[55]   Sarret G, Vangronsveld J, Manceau A, *et al.* Accumulation forms of Zn and Pb in *Phaseolus vulgaris* in the presence and absence of EDTA. Environ Sci Technol 2001; 35: 2854-9.

[56]   Sharma NC, Gardea-Torresdey JL, Parsons J, Sahi SV. Chemical speciation and cellular deposition of lead in *Sesbania drummondii*. Environ Toxicol Chem 2004; 23: 2068-73.

[57]   Meyers DE, Kopittke PM, Auchterlonie GJ, Webb RI. Characterisation of lead precipitate following uptake by roots of *Brassica juncea*. Environ Toxicol Chem 2009; 28: 2250-4.

[58]   Jarvis MD, Leung DWM. Chelated lead transport in *Chamaecytisus proliferus* (L.f.) link ssp. *proliferus* var. *palmensis* (H. Christ): an ultrastructural study. Plant Sci 2001; 161: 433-41.

[59]   Epstein AL, Gussman CD, Blaylock MJ, *et al.* EDTA and Pb-EDTA accumulation in *Brassica juncea* grown in Pb-amended soil. Plant Soil 1999; 208: 87-94.

[60]   Jarvis MD, Leung DWM. Chelated lead transport in *Pinus radiata*: an ultrastructural study. Environ Exp Bot 2002; 48: 21-32.

[61]   Zheng LJ, Peer T, Seybold V, Luetz-Meindl U. Pb-induced ultrastructural alternations and subcellular localization of Pb in two species of *Lespedeza* by TEM-coupled electron energy loss spectroscopy. Environ Exp Bot 2012; 77: 196-206.

[62] Jadia CD, Fulekar MH. Phytoremediation of heavy metals: recent techniques. Afri J Biotechnol 2009; 8: 921-8.

[63] Butcher DJ. Phytoremediation of lead in soil: recent applications and future prospects. Appl Spectrosc Rev 2009; 44: 123-39.

[64] Saifullah, Meers E, Qadir M, *et al.* EDTA-assisted Pb phytoextraction. Chemosphere 2009; 74: 1279-91.

[65] Blaylock MJ, Salt DE, Dushenkov S, *et al.* Enhanced accumulation of Pb in Indian mustard by soil-applied chelating agents. Environ Sci Technol. 1997; 31: 860-5.

[66] Epelde L, Hernandez-Allica J, Becerril JM, Blanco F, Garbisu C. Effects of chelates on plants and soil microbial community: Comparison of EDTA and EDDS for lead phytoextraction. Sci Total Environ 2008; 401: 21-8.

[67] Huang JW, Cunningham SD. Lead phytoextraction: Species variation in lead uptake and translocation. New Phytol 1996; 134: 75-84.

[68] Freitas EVD, do Nascimento CWA. The use of NTA for lead phytoextraction from soil from a battery recycling site. J Hazard Mat 2009; 171: 833-7.

[69] Cui, YS, Wang QR, Dong YT, Li HF, Christie P. Enhanced uptake of soil Pb and Zn by Indian mustard and winter wheat following combined soil application of elemental sulphur and EDTA. Plant Soil 2004; 261: 181-8.

[70] Begonia MT, Begonia GB, Ighoavodha M, Gilliard D. Lead accumulation by tall fescue (*Festuca arundinacea* Schreb.) grown on a lead-contaminated soil. Int J Environ Res Public Health 2005; 2: 228-33.

[71] López, M.L., *et al.*, Gibberellic acid, kinetin, and the mixture indole-3-acetic acid-kinetin assisted with EDTA-induced lead hyperaccumulation in alfalfa plants. Environmental Science & Technology, 2007. 41:8165-70.

[72] Cao A, Carucci A, Lai T, La Colla P, Tamburini E. Effect of biodegradable chelating agents on heavy metals phytoextraction with *Mirabilis jalapa* and on its associated bacteria. European J Soil Biol 2007; 43: 200-6.

[73] Hassan M, Sighicelli M, Lai A, *et al.* Studying the enhanced phytoremediation of lead contaminated soils *via* laser induced breakdown spectroscopy. Spectrochim Acta B 2008; 63: 1225-9.

[74] Liu D, Islam E, Ma JS, *et al.* Optimization of chelator-assisted phytoextraction, using EDTA, lead and *Sedum alfredii* Hance as a model system. Bullet Environ Contaminat Toxicol 2008; 81: 30-5.

[75] McDonald, Phytoremediation of lead-contaminated soil using *Typha latifolia* (Broadleaf cattail). Master thesis 2006.

[76] Li YL, Liu YG, Liu JL, Zeng GM, Li X. Effects of EDTA on lead uptake by *Typha orientalis Presl*: A new lead-accumulating species in southern China. Bullet Environ Contaminat Toxicol 2008; 81: 36-41.

[77] Cui S, Zhou QX, Wei SH, Zhang W, Cao L, Ren L.P. Effects of exogenous chelators on phytoavailability and toxicity of Pb in *Zinnia elegans* Jacq. J Hazard Mat 2007; 146: 341-6.

[78] Chen YH, Li XD, Shen ZG. Leaching and uptake of heavy metals by ten different species of plants during an EDTA-assisted phytoextraction process. Chemosphere 2004; 57: 187-96.

CHAPTER 3

Effect of Nitric Oxide Donors on Metal Toxicity in Plants

David W.M. Leung[*]

School of Biological Sciences, University of Canterbury, Private Bag 4800, Christchurch 8140, New Zealand

Abstract: Nitric oxide (NO) has been shown to be an important signaling molecule in mammalian and plant physiology. The notion that exogenous application of NO in the form of a solution-based NO donor, for example, sodium nitroprusside (SNP), can counteract the toxicity of heavy metals in plants has been supported experimentally in many studies in the past decade. However, some recent studies also appeared to have casted doubts about this. Moreover, there does not appear to have been any assessment of the practical or agricultural significance of applying NO exogenously for ameliorating heavy metal toxicity in plants, particularly during postgerminative seedling growth. The main features of the relevant studies were examined critically. The issues discussed in relation to the studies of applying NO and heavy metal treatment of seedlings during postgerminative growth might also be relevant to studies at other plant growth and developmental stages. It is concluded that the agricultural significance of exogenous application of NO to alleviate heavy metal toxicity in plants remains to be established.

Keywords: Abiotic stress, alleviation of metal toxicity, antioxidants, *Arabidopsis thaliana*, catalase, cPITO, crop protection, environmental pollution, morphological alterations, nitric oxide, oxidative stress, peroxidase, postgerminative seedling growth, pretreatment, reactive oxygen species (ROS), root elongation, root growth inhibition, sodium nitroprusside (SNP), yellow lupin.

INTRODUCTION

Modern anthropogenic activities have unwittingly been accompanied by environmental pollution [1]. For example, metalliferous mining, car exhausts, use of fertilizers for agriculture, horticulture and forestry *etc*. have contributed to increasing heavy metal pollution in the environment [1, 2]. Toxicity or stress in

*Address correspondence to David W.M. Leung: School of Biological Sciences, University of Canterbury, Private Bag 4800, Christchurch 8140, New Zealand; Tel: 64 3 3642650; Fax: 64 3 3642590; E-mail: david.leung@canterbury.ac.nz

plants could be induced by heavy metals dependent on concentrations [3]. Research interest in this area is a sub-category of and is closely related to the studies on the biology of abiotic stress in plants [3]. Abiotic stress in crop plants could severely limit their agricultural productivity and is the primary threat to crop failure worldwide [4]. Research into this problem is of contemporary agricultural significance and central to the well-beings of modern economies globally.

There is another practical impetus in studying the relationship between plants and heavy metal contaminants in the environment. It has been proposed that some plants could be deployed to remove heavy metals from contaminated soil and water [5]. This plant-based environmental biotechnology or phytomanagement strategy is called phytoremediation. An important prerequisite for choosing the suitable plants for phytoremediation is a better understanding of toxic metal-triggered injurious alterations as well as the limits and mechanisms of toxic metal tolerance in plants. Many experimental studies have been carried out using soils, or hydroponic solutions, or even under plant tissue culture conditions artificially contaminated with a heavy metal. Symptoms of metal toxicity in plants can be manifested as morphological alterations such as inhibition of root growth. Therefore toxic effects of heavy metals result in perturbations to the underlying physiological, cellular, biochemical and molecular mechanisms associated with normal plant growth and development.

It was first reported that sodium nitroprusside (SNP), a nitric oxide (NO) donor, supplied as an incubation solution for pretreatment of yellow lupin seedlings before exposure to Pb or Cd had prevented metal-induced inhibition of root growth [6]. Recently, there are many research papers showing results in support for a role of exogenously applied nitric oxide (NO) in alleviating heavy metal toxicity in plants. It has also been proposed that modulation by SNP of metal-induced oxidative stress is a plausible mechanism involved (for example, [7, 8]). However, recently a concern has also been raised regarding results showing that endogenous NO might contribute to Cd toxicity in plants [9, 10]. Furthermore, a different group working on the effect of a supply of SNP on Al toxicity in plants has recently shown that simultaneous application of SNP and Al led to more inhibition of root growth than Al treatment alone [11]. Taken together, the story about the relationship between an exogenous supply of NO and heavy metal toxicity in plants must now be considered as an evolving one. The aim of this review is to examine more closely the diversity of the findings in this kind of investigation. Additionally, the practical significance of the protective effect of an exogenous supply of NO on heavy metal toxicity in plants reported thus far will be discussed.

DOSE-DEPENDENT EFFECT OF SODIUM NITROPRUSSIDE (SNP)

Primary root length of 3-d-old tomato seedlings incubated in 50 μM SNP for 5 days was the same as that of the control [12]. Compared to the control, higher concentrations of SNP inhibited root elongation of tomato seedlings in a dose-dependent manner. Similarly, exogenous application of SNP, dependent on its concentration, affected root elongation in 2-d-old wheat seedlings [13] and 5-d-old rice seedlings [14]. In addition, 1 mM cPITO, a scavenger of NO, promoted an increase in primary root length of tomato seedlings compared to the control, suggesting that endogenous NO was normally involved in primary root elongation. Taken together, these findings obtained in the absence of heavy metals have implications for studies of artificially supplying NO *via* SNP for ameliorating toxic effects of heavy metals on root elongation. For these studies, it is a pre-requisite to determine, from a dose-dependent growth curve, a SNP concentration that has no inhibitory effect on root elongation.

STUDIES LINKING TOXICITY OF HEAVY METALS TO INHIBITION OF ROOT GROWTH

The first visible sign of successful seed germination is emergence of the radicle or the embryonic root through the seed coat and surrounding structures associated with a seed. This marks the beginning of early postgerminative seedling growth which is an important phase of crop plant production. During this growth phase, the first main developmental process is associated with the need for the successful establishment of a functional primary root that can provide anchorage of the young seedling to the growth substrate (usually the soil). Additionally, the growing primary root mainly increases in length needed to explore the growth substrate far and wide for water and mineral nutrients. Inevitably the primary root could encounter biotic and/or abiotic stesss factors such as toxic levels of environmental heavy metal contaminants. These have been shown to be capable of inhibiting root growth (Table **1**). Protection of the primary root from the harmful effects of heavy metals is, therefore, of considerable agricultural interest and significance.

Table 1: Investigations into toxic effects of heavy metals in the presence or absence of SNP on inhibition of root growth of various plant species

Plant species	Age (d, unless indicated otherwise)	Treatment with Metal (µM), in the absence or presence of NO donor (µM)	Duration of Metal Treatment (h, unless indicated otherwise)	Root length (% of control)*		References
				Metal treatment alone	Metal + exogenous NO donor	
Cassia tora	3	AlCl₃ (10), SNP (400)	24	42	50	[15]
Vigna umbellata (rice bean)	3	AlCl₃ (25), SNP (400)	24	65	33	[11]
Oryza sativa (rice)	4	AlCl₃ (75), SNP (25)	24	53	89	[16]
Secale cereale (rye) *Triticum aestivum* (wheat)	4	AlCl₃ (50), SNP (1000)	24	43.7 10	60 19	[17]
Arabidopsis thaliana	7	Pb(NO₃)₂ (100), SNP (500)	7 d	29	37	[8]
Triticum aestivum (wheat)	4	Pb(NO₃)₂ (4000), SNP (25)	6 d	77	84 (NS)	[18]
Lupinus luteus (yellow lupin)	Several d-old (Not specified)	PbCl₂ (1500) and CdCl₂ (50), SNP (10)	48	37	53	[6]
Oryza sativa (rice)	5	As (50), SNP (50)	24	63	78	[14]
Medicago truncatulq	Several d-old seedlings (not specified)	CdSO₄ (50), SNP (100)	48	58	63	[19]

Table 1: contd....

Phaseolus vulgaris (bean)	5	$NiCl_2$ (200), SNP (300)	4 d	50	76	[20]
Lycopersicon esculentum (tomato)	5 wk	$CuCl_2$ (50), SNP (100)	8 d	50 (dry root weight)	100	[21]
(*Triticum aestivum*) (wheat)	4	$CuSO_4$ (50) or $ZnSO_4$ (50), SNP (500)	unclear	30 35 (Dry root weight)	48 50	[22]
Oryza sativa (rice)	4 wk	$CdCl_2$ (200), SNP (50)	10 d	78	91	[23]
Oryza sativa (rice)	9	$Cd\ Cl_2$ (100), SNP (200)	7 d	23	23	[24]
Hibiscus moscheutos	Young seedlings, age not specified	$Al\ Cl_3$ (100), SNP (100)	24	24	58	[25]
Triticum aestivum (wheat) *Phaseolus vulgaris* (bean)	Young seedlings; Unknown age	$ZnSO_4$ (21.6), SNP (100)	21	50 58 (root biomass)	79 74	[26]

*(Root length of control - Root length of seedlings exposed to a metal treatment or metal+SNP)/root length of control x 100%; the root length measurements used in the calculations here were estimates extracted from the given references.

Interestingly, a majority of studies have obtained evidence for a role of an exogenous NO supply in alleviating toxicity of heavy metals from experiments on seedlings at the early postgerminative growth stage (Table **1**). Typically, seeds were first germinated on paper wetted with water. Then treatment with toxic levels of a heavy metal with or without an exogenous NO supply often began when the seedlings were of several days old and often when the elongating primary root was the only growing plant organ with the most observable/measureable morphological changes (in terms of root length or colour changes *etc.*). The ease to set up the treatments and control of the experimental materials have no doubt contributed to the popularity of this stage of plant development or crop establishment for most studies on the problem of heavy metal toxicity and protection.

Having shown a relationship between an exogenous supply of NO and heavy metal-triggered inhibition of root growth in seedlings, the main focus of most studies has been justifiably to unravel the underlying mechanism(s) involved. For example, it might be possible that exogenous application of SNP had a protective effect on *Arabidopsis thaliana* seedlings exposed to toxic levels of Pb because the exogenous NO supply could evoke an avoidance mechanism regarding Pb uptake in *A. thaliana* seedlings [8]. To evaluate this possibility, it was important to determine heavy metal uptake and accumulation in plants exposed to heavy metal in the presence or absence of an exogenous supply of NO although this was reported only in some studies (Table **2**). In the *A. thaliana* study, the SNP treatment had no effect on Pb uptake and therefore the results obtained did not show support for the metal uptake avoidance possibility [8]. Instead, alleviation by the SNP treatment of root growth of *A. thaliana* seedlings under otherwise Pb stress was correlated with a raft of increases in the activities of enzymes such as catalase and peroxidase associated with antioxidative defence [8] as well as prevention of increased oxidative stress associated with exposure of plant tissues to toxic levels of heavy metals. This result was consistent with those of many other similar studies [7]. Besides, some recent studies showed that exogenous application of SNP was able to increase pectin and hemicellulose contents in the cell walls of rice seedling roots exposed to $CdCl_2$ This was correlated with more accumulation of Cd^{2+} in the cell walls [23], suggesting that SNP could protect plant roots from metal stress by promoting increased sequestration of heavy metals in the cell wall, assuming there were less targets susceptible to toxicity of heavy metals in the cell wall than in the cytoplasm.

Table 2: Determination of NO and metal contents in studies concerning the protective effect of SNP against metal toxicity on plant root growth

Plant species	Effect of SNP on metal content in roots	NO content in roots in response to metal or SNP+metal treatment	References
Cassia tora	20% less Al	n.d.*	[15]
Vigna umbellata (rice bean)	Increased accumulation of Al in root apices	The same levels of NO in control and the Al (25 µM) treatment; In absence of Al, applied SNP (400 µM) resulted in a higher level of NO than control (no Al, no SNP); the same levels of NO in SNP+Al and SNP alone; this SNP conc. alone had no inhibitory effect on root growth	[11]

Table 2: contd….

Oryza sativa (rice)	Reduced Al content in root apices	n.d.	[16]
Secale cereale (rye) & *Triticum aestivum* (wheat)	Decreased Al uptake by about 60%	Increased by about 70%	[17]
Arabidopsis thaliana	No effect	n.d.	[8]
Triticum aestivum (wheat)	n.d.	n.d.	[18]
Lupinus luteus (yellow lupin)	n.d	n.d.	[6]
Oryza sativa (rice)	As accumulation reduced by about 40% in the root	As alone reduced endogenous NO production	[14]
Medicago truncatula	Reduced Cd accumulation	Cd Reduced NO accumulation	[19]
Phaseolus vulgaris (bean)	Increased Ni content of the root	n.d.	[20]
Lycopersicon esculentum (tomato)	No effect on Cu content in roots	n.d.	[21]
(*Triticum aestivum*) (wheat)	n.d.	n.d.	[22]
Oryza sativa (rice)	Increased Cd accumulation in the root and shoot but decreased in leaf	Transient increase and lower at the end of treatment (24 h) than control	[23]
Oryza sativa (rice)	n.d.	Cd reduced NO level; SNP partially restored NO level	[24]
Hibiscus moscheutos	n.d.	Al reduced NO level compared to control; SNP restored back to control level	[25]
Triticum aestivum (wheat) & *Phaseolus vulgaris* (bean)	Decreased Zn uptake in wheat root but increase in bean	n.d.	[26]

*n.d.= not determined.

It is pertinent to note that the duration of most, if not all, of the experiments involving application of SNP was for a day or two. Therefore, it is difficult, based on these short-term experiments, to ascertain if any protection from heavy metal stress correlated with the SNP treatment would be a long-lasting benefit for further plant growth and development or not. This is an important issue for protection of crops from potential harms of environmental heavy metal contaminants and should be clarified in future studies.

THE MAGNITUDE OF PROTECTION BY SNP

There are obvious practical challenges if NO, a gas, is to be applied in the field for agricultural benefits. Although a NO donor solution such as SNP could be used to counteract the harmful effects of heavy metals, there are some concerns regarding the use of NO donors in plant studies based on their chemistry and metabolic effects [27]. What seems to be clear is that the strength of alleviation of heavy metal stress by an exogenous supply of NO in various studies has not been taken into account, particularly from the perspective of the relative agricultural importance of this potential crop plant protection approach.

A survey of the results obtained so far has revealed that exogenous application of SNP in relation to treatment of plants with toxic levels of heavy metals has mainly resulted in partial or incomplete protection of root elongation from toxicity of heavy metals (Table **1**). Often there was only a small % improvement of root growth when seedlings were treated with SNP + a heavy metal compared with those treated with the heavy metal alone in most studies (Table **1**). This might not be too surprising as regulation of root growth under other abiotic stress conditions often involves a multitude of interacting factors and signaling mechanisms [28]. Metal toxicity on root growth might impact on these networks of factors and signaling mechanisms as well. Besides, NO might just be one of the signaling molecules involved in root growth. Nevertheless, it seems that exogenous application of NO might be of academic interest only and studies of the underlying mechanism involved have generally contributed to a better understanding of the signaling function of NO in plant growth and abiotic stress physiology. Better and more practical protection measures for crops from the harms of environmental heavy metal contaminants remain elusive at present.

CONTRADICTORY FINDINGS

It was shown that treatment of rice seedlings for 7 d with 100 μM $CdCl_2$ resulted in 77% reduction in root length [24]. Contrary to the findings of many similar studies, application of 200 μM SNP was not able to counteract the toxic effect of $CdCl_2$ on root elongation (Table **1**). In another study, application of 25 μM SNP also could not protect wheat seedlings from the treatment with 4 mM $Pb(NO_3)_2$ which resulted in 23% inhibition of root elongation ([18]; Table **1**). Interestingly, the SNP treatment was beneficial for shoot growth of the seedlings under the same heavy metal stress condition [18].

The possibility that root elongation under heavy metal stress was significantly worse off in the presence of SNP was recently demonstrated [11]. When *Vigna umbellata* seedlings were treated with 25 μM $AlCl_3$ for 24 h root elongation was inhibited by 35% but in the treatment of 25 μM $AlCl_3$ + 400 μM SNP root elongation was inhibited by 67% (Table **1**). The more severe reduction in root length was not due to the primary action of SNP as the level SNP used alone did not inhibit root elongation compared to the control in the absence of the toxic contaminant. In contrast, an earlier study using the same levels of $AlCl_3$, SNP and treatment duration but on another plant (*Cassia tora*) showed that there was a small (8%) but statistically significant ($p < 0.05$) improvement in root length of the seedlings treated with $AlCl_3$ + SNP over the treatment with $AlCl_3$ alone [15]. The different findings could be due to different plants being used in the different studies. Nevertheless, it is of interest to elucidate more clearly the basis for this in future studies.

STATUS OF ENDOGENOUS NITRIC OXIDE (NO) LEVEL

There is evidence that endogenous NO normally (that is, in the absence of heavy metals) plays an essential role in regulating root elongation in tomato seedlings as application of 1 mM cPITO, a NO-specific scavenger led to an increase in root length in tomato seedlings compared to the water control [12]. However, application of 250 μM cPITO or 500 μM of a mammalian nitric oxide synthase inhibitor (L-NAME) resulted in a reduction of primary root length in *A. thaliana* seedlings [29]. While there is no common consensus regarding the role of endogenous NO in the absence of heavy metals, treatment of 3-week-old *A. thaliana* seedlings with 200 μM CdSO$_4$ for 7 h resulted in more NO production than the control [9]. Treatment with cPITO or L-NAME) was found to partially overcome inhibition of root elongation in the presence of 300 μM CdSO$_4$ [29]. Therefore, lowering rather than augmentation of endogenous NO level was correlated with root growth improvement in the presence of Cd. These were the first studies that casted doubts about the possible protective effect of an exogenous supply of NO on heavy metal stress in plants even though both studies did not pretreat or simultaneously supply seedlings with a NO donor and a toxic metal.

Artificially augmenting NO *via* application of SNP and exposure to heavy metals could change the NO levels of the roots in the seedlings under investigation. Could a diversity of findings from these studies be somehow related to the varying levels of NO in the plants during experimentation? Therefore, to have a better and more precise interpretation of the results obtained from this kind of investigations, it is important to monitor the levels of NO in the plants in the various treatments. Unfortunately, many studies showing the protection from the harmful effect of a heavy metal on the seedling root length with an exogenous supply of NO did not determine endogenous NO level in the root (Table **2**). Of those studies that did determine endogenous NO levels, it was shown that at the end of a 24-h treatment the NO levels were reduced in response to Al [25], Cd [19, 24], and As [14]. At least in these studies, it would seem reasonable to conclude that an exogenous supply of NO might be able to replace the reduced level of endogenous NO brought about by the exposure to a heavy metal. This could then be correlated with the protection of the seedling root exposed to the heavy metal.

There were other studies showing that Al treatment had no effect on endogenous NO level in rice bean [11], but Al was shown to increase NO level in rye and wheat [17]. In a time-course study, Cd treatment was shown to bring about a transient increase in NO level before a decrease in rice [23]. In these studies, it would be difficult to simply explain that an exogenous supply of NO could alleviate the toxic effect of a heavy metal by way of supplementing a metal-induced deficiency in endogenous level of NO.

CONCLUSIONS

Environmental heavy metal pollution could have impact on plant growth and development beyond early postgerminative seedling growth. It is also clear that NO seems to be a signal molecule involved in many physiological and developmental processes and root growth is just one of them. An analysis of a novel cadmium resistant mutant showed that Cd resistance is linked to flowering time and seed size determination [30]. Pharmacological evidence from exogenous application of Cd showed that Cd could promote early flowering in *A. thaliana* [30] but delay flowering in chickpeas [31]. In the former study, application of SNP and/or a scavenger of NO showed that the Cd-triggered premature flowering was linked to a NO signaling effect. In the latter study, the underlying mechanism was not investigated. All the issues identified so far in the diverse findings related to the protective effect of SNP on heavy metal-triggered root growth inhibition might also be relevant to other studies on different facets of plant growth and development involving the application of an exogenous supply of a NO donor. Additionally, the practical significance or implications from a better understanding of the role of NO in plant physiological processes under heavy metal stress should not be neglected. The quest for improvement of crop plant productivity using NO donors or whatever innovative approaches in the face of increasing environmental heavy metal pollution is to be supported with more quality investigations.

ACKNOWLEDGEMENTS

Declared none.

CONFLICT OF INTEREST

The author(s) confirm that this chapter content has no conflict of interest.

REFERENCES

[1]   Nagajyoti PC, Lee KD, Sreekanth TVM. Heavy metals, occurrence and toxicity for plants: a review. Eviron Chem Lett 2010; 8: 199-216.

[2]   Rascio N, Navari-Izzo F. Heavy metal hyperaccumulating plants: How and why they do it? And what makes them so interesting? Plant Sci 2011; 180: 169-81.

[3]   Siddiqui MH, Al-Whaibi MH, Basalah MO. Role of nitric oxide in tolerance of plants to abiotic stress. Protoplasma 2011; 248: 447-55.

[4]   Witcombe JR, Hollington PA, Howarth CJ, Reader S, Steele KA. Breeding for abiotic stresses for sustainable agriculture. Phil Trans R Soc B 2008; 363: 703-16.

[5]   Pilons-Smits E. Phytoremediation. Annu Rev Plant Biol 2005; 56: 15-39.

[6]   Kopyra M, Gwozdz EA. Nitric oxide stimulates seed germination and counteracts the inhibitory effect of heavy metals and salinity on root growth of *Lupinus luteus*. Plant Physiol Biochem 2003; 41: 1011-7.

[7]   Xiong J, Fu G, Tao L, Zhu C. Roles of nitric oxide in alleviating heavy metal toxicity in plants. Arch Biochem Biophys 2010; 497: 13-20.

[8]   Phang IC, Leung DWM, Taylor HH, Burritt DJ. The protective effect of sodium nitroprusside (SNP) treatment on *Arabidopsis thaliana* seedlings exposed to toxic level of Pb is not linked to avoidance of Pb uptake. Ecotoxicol Environ Safety 2001; 74: 1310-5.

[9]   Besson-Bard A., Wendehenne D. NO contributes to cadmium toxicity in *Arabidopsis thaliana* by mediating an iron deprivation response. Plant Signal Behav 2009; 4: 252-4.

[10]   Arasimowicz-Jelonek M, Floryszak-Wieczorek MJ, Gwozdz EA. The message of nitric oxide in cadmium challenged plants. Plant Sci 2011a; 181: 612-20.

[11]   Zhou Y, Xu XY, Chen LQ, Yang JL, Zheng SJ. Nitric oxide exacerbates Al-induced inhibition of root elongation in rice bean by affecting cell wall and plasma membrane properties. Phytochem 2012; 76: 46-51.

[12]   Correa-Aragunde N, Graziano M, Lamattina L. Nitric oxide plays a central role in determining lateral root development in tomato. Planta 2004; 218: 900-5.

[13]   Groppa MD, Rosales EP, Iannone MF, Benavides MP. Nitric oxide, polyamimes and Cd-induced phytotoxicity in wheat roots. Phytochem 2008; 69: 2609-15.

[14]   Singh HP, Kaur S, Batish DR, Sharma VP, Sharma N, Kohli RK. Nitric oxide alleviates arsenic toxicity by reducing oxidative damage in the roots of *Oryza sativa* (rice). Nitric Oxide-Biol Chem 2009; 20: 289-97.

[15]   Wang YS, Yang ZM. Nitric oxide reduces aluminum toxicity by preventing oxidative stress in the roots of *Cassia tora* L. Plant Cell Physiol 2005; 46: 1915-23.

[16]   Zhang Z, Wang H, Wang X, Bi Y. Nitric oxide enhances aluminum tolerance by affecting cell wall polysaccharides in rice roots. Plant Cell Rep 2011; 30: 1701-11.

[17]   He HY, He LF, Gu MH, Li XF. Nitric oxide improves aluminum tolerance by regulating hormonal equilibrium in the root apices of rye and wheat. Plant Sci 2012; 183: 123-130.

[18]   Yang Y, Wei X, Lu J, You J, Wang W, Shi R. Lead-induced phytotoxicity mechanism in seed germination and seedling growth of wheat (*Triticum aestivum* L.) Ecotoxicol Environ Safety 2010; 73: 1982-7.

[19]   Xu J, Wang W, Yin H, Liu X, Sun H, Mi Q. Exogenous nitric oxide improves antioxidative capacity and reduces auxin degradation in roots of *Medicago truncatula* seedlings under cadmium stress. Plant Soil 2010; 326: 321-30.

[20]   MihailovicN, Drazic G. Incomplete alleviation of nickel toxicity in bean by nitric oxide supplementation. Plant Soil Environ 2011; 57: 396-01.

[21]   Zhang Y, Han X, Chen X, Jin H, Cui X. Exogenous nitric oxide on antioxidative system and ATPase activities from tomato seedlings under copper stress. Sci Hortic 2009; 123: 217-23.

[22]   Gil'vanova IR, Enikeev AR, Stepanov SY, Rakhmankulova ZF. Involvement of salicylic acid and nitric oxide in protective reactions of wheat under the influence of heavy metals. Applied Biochem Microbiol 2012; 48: 90-4.

[23]   Xiong J, An L, Lu H, Zhu C. Exogenous nitric oxide enhances cadmium tolerance of rice by increasing pectin and hemicellulose contents in root cell wall. Planta 2009a; 230: 755-65.

[24]   Xiong J, Lu H, Lu K, Duan Y, An L, Zhu C. Cadmium decreases crown root number by decreasing endogenous nitric oxide, which is indispensable for crown root primordia initiation in rice seedlings. Planta 2009b; 230: 599-610.

[25]   Tian QY, Sun DH, Zhao NG, Zhang WH. Inhibition of nitric oxide synthase (NOS) underlies aluminum-induced inhibition of root elongation in *Hibiscus moscheutos*. New Phytologist 2007; 174: 322-31.

[26]   Abdel-Kader DZE. Role of nitric oxide, glutathione and sulfhydryl groups in zinc homeostasis in plants. Am J Plant Physiol 2007; 2: 59-75.

[27]   Arasimowicz-Jelonek M, Floryszak-Wieczorek MJ, Kosmala A. Are nitric oxide donors a valuable tool to study the functional role of nitric oxide in plant metabolism? Plant Biol 2011b; 13: 747-56.

[28]   Potters G, Pasternak TP, Guisez Y, Jansen MAK. Different stresses, similar morphogenetic responses: integrating a plethora of pathways. Plant Cell Environ 2009; 32: 158-69.

[29]   Besson-Bard A, Gravot A, Richard P, *et al*. Nitric oxide contributes to cadmium toxicity in *Arabidopsis* by promoting cadmium accumulation in roots and by upregulating genes related to iron uptake. Plant Physiol 2009; 149: 1302-15.

[30]   Wang Y, Zhong K, Jiang L, Sun, Y. Characterization of an *Arabidopsis* cadmium-resistant mutant *cdr3-1D* reveals a link between heavy metal resistance as well as seed development and flowering. Planta 2011; 233: 697-706.

[31]   Wani PA, Khan MS, Zaidi A. Impact of heavy metal toxicity on plant growth, symbiosis, seed yield and nitrogen and metal uptake in chickpea. Aust J Exp Agri 2007; 47: 712-20.

Send Orders for Reprints to reprints@benthamscience.net

CHAPTER 4

Metal Hyperaccumulating Ferns: Progress and Future Prospects

Sarita Tiwari, Bijaya K. Sarangi[*], Pulavarty Anusha and Ram A. Pandey

Environmental Biotechnology Division, CSIR-National Environmental Engineering Research Institute, Nehru Marg, Nagpur - 440020, India

Abstract: Environmental exposure to heavy metals such as arsenic and chromium has become a serious problem worldwide. The recent advances in the knowledge of metal hyperaccumulation by some plant species have recognized phytoextraction as a prospective proposition using hyperaccumulator plants. This paper enumerates the progress in phytoextraction of metals using fern species with emphasis on the arsenic hyperaccumulator *Pteris vittata*. The scopes of R&D for value addition in ferns for efficient phytoremediation application have also been discussed.

Keywords: Aluminum (Al), aquatic ferns, *Athyrium yokoscense*, *Azolla sliculoides*, arsenic (As), biosphere, cadmium (Cd), chromium (Cr), copper (Cu), ferns, glutathione (GSH), *Holcus lantus*, hyperaccumulator, lead (Pb), mercury (Hg), metal transport, nickel (Ni), *Pityrogramma calomelanos*, *Pteris vittata*, zinc (Zn).

FERNS, THEIR SPECIFICITY, SYSTEMS FOR METAL ACCUMULATION

Toxic metal pollution of the biosphere has intensified rapidly since the onset of the industrial revolution, posing major environmental and health problems. Heavy metals have densities higher than 5 g cm^{-3}, and include cadmium (Cd), mercury (Hg), lead (Pb), copper (Cu), and zinc (Zn), among others. These are among the most important pollutants causing worldwide environmental contamination and human health problems. Besides these, there are other metals/metalloids like arsenic (As), chromium (Cr), Aluminum (Al) which cause severe concerns due to

*Address correspondence to Bijaya K. Sarangi: Environmental Biotechnology Division, CSIR-National Environmental Engineering Research Institute, Nehru Marg, Nagpur - 440020, India; Tel: +917122249886; Ext: 415; Cell: +919421706693; Fax: +917122249900; E-mail: bk_sarangi@neeri.res.in

their toxicity and some of them are topical. The fern species exhibit phenotypic variation in response to metal ions and some species like *Pteris vittata* have evolved as arsenic (As) hyperaccumulators. Studies of these plants help to understand As toxicity, the way in which plants have evolved As resistance, and scope for application in phytoremediation of As pollution.

Ferns are adapted to extreme environments. Recent molecular phylogenetic analyses indicated that ferns have undergone recent adaptive radiations [1]. Several fern species of Actiniopteridaceae, Sinopteridaceae, Pteridaceae and Selaginellaceae have been documented to be distributed in rock outcrops in the tropics and to exhibit desiccation tolerance [2]. The genus *Acrostichum aureum*, the halophytic ferns, grow in association with mangrove vegetation in the tropics. They could accumulate the cyclitol d-1-O-methyl-muco-inositol [3], a cytoplasmic compatible solute, in response to increasing salinity. The primitive fern, *Osmunda cinnamomea*, is tolerant of copper, cadmium and zinc [4]. *Athyrium yokoscense* accumulates lead in its tissues, particularly in the roots [5], and the gametophytes of this species also exhibit lead tolerance and accumulation [6]. The ability of ferns to grow on metal-contaminated soils has been recognized although they were not discussed much before the report on As-tolerant *Pteris vittata* [7]. In other studies, it was found that *Asplenium adulterium* was an indicator of nickel [8], while *Pellaea calomelanos* and *Chelianthes hirta* were found on copper and occasionally nickel soils [9]. The fern, *Asplenium septentrionale* (L.) was found in old lead and copper workings in North and Central Wales [10]. Ferns were also recorded on serpentine soils high in nickel and chromium [11]. Apart from terrestrial ferns, some aquatic ferns such as *Azolla sliculoides* are able to take up large concentrations of heavy metals and accumulated in the shoots [12]. Other ferns with metal accumulating capabilities include *Salvinia natans* for copper [13], *S. molesta*, *Azolla pinnath*, *Marsilea minuta* for cadmium [14], and *S. minima* for chromium [15]. Other than heavy metals, ferns have also been known to concentrate large quantities of trace elements in their tissues [16].

A range of fern species (45) and their allies, *Equisetum* (5), *Selaginella* (2) species and *Psilotum nudum* were screened [17] for their ability to hyperaccumulate As and develop a phylogenetic understanding of this phenomenon. A number of varieties (5) of a known As hyperaccumulator *Pteris cretica* were additionally included in this study. This was the first report showing the members of the *Pteris* genus, *P. straminea* and *P. tremula*, do not hyperaccumulate As. Twelve species of ferns were screened [18] for their ability to tolerate and hyperaccumulate As. Ferns were exposed to 50 or 100 mg As L^{-1} for 7 and 14 days using hydroponic (soil free) experiments. The fronds and roots were analysed for As, selected macronutrients (K, Ca, Mg, P and S) and micronutrients (Al, Fe, Cu and Zn). Five fern species (*Asplenium aethiopicum, A. australasicum, A. bulbiferum, Doodia heterophylla and Microlepia strigosa*) were found to be sensitive to As. However, only *A. Australasicum* and *A. Bulbiferum* could hyperaccumulate As up to 1240 and 2630 µg As g^{-1} dry weight (dw), respectively, in their fronds after 7 days at 100 mg As L^{-1}. This is the first known report of ferns that are sensitive to As but yet are As hyperaccumulators. All As-tolerant ferns (*Adiantum capillus-veneris, Pteris cretica var. Albolineata, P. cretica* var. Wimsetti and *P. umbrosa)* were from the Pteridaceae family. *P. Cretica and P. Umbrosa* accumulated the majority of As in their fronds (up to 3090 µg As g^{-1} dw) compared to the roots (up to 760 µg As g^{-1} dw). In contrast, *A. capillusveneris* accumulated the majority of As in the roots (1190 µg As g^{-1} dw) compared to the fronds (370 µg As g^{-1} dw). The only non-*Pteris* fern to exhibit this ability is *Pityrogramma calomelanos* [19]. Its fronds were able to accumulate 2760 to 8350 mg As kg^{-1} when grown in soil containing 135 to 510 mg As kg^{-1}. It was suggested that *Pityrogramma* has the ability to effectively reduce soil arsenic concentrations and is a better phytoextraction candidate than *P. vittata* because it appeared able to grow better in the arsenic-contaminated soils.

PLANT RESPONSE TO METAL STRESS

Plants have developed three basic strategies for growing on contaminated and metalliferous soils [20]. These plants can be grouped under different categories on the basis of their morphological, physiological and biochemical mechanisms of metal tolerance.

1. Metal indicators: These plants accumulate metals in their above-ground tissues and the metal levels in the tissues of these plants generally reflect metal levels in the soil.

2. Metal avoiders: These plants can detect the occurrence of the metal stressors and mostly develop morphological adaptations to prevent entry of the toxic metals into the plant system and thereby remain unaffected by them.

3. Metal toleraters: These species are devoid of mechanisms to prevent entry of metals into the cellular milieu. Their tolerance to the stress is limited. Above a critical concentration of the metal ions the biochemical and physiological mechanisms cease to be functional and the plants die.

4. Metal excluders: These plants effectively prevent metal ions from entering into their aerial parts over a broad range of metal concentrations in the soil. However, they can still contain large amounts of metals in their roots.

5. Accumulators & hyperaccumulators: These are unique plant species (hyperaccumulators) as they can concentrate metals in their above-ground tissues to levels far exceeding those present in the soil or in the nonaccumulating species growing nearby. In the presence of high concentrations of the metal ions inside the cytosol, they are capable of carrying out biochemical and physiological activities and remain alive. The bioaccumulation factor (BF) of the hyperaccumulators are characterized as having a plant to soil ratio of greater than one [21]. It has been proposed that a plant with more than 0.1% of Ni, Co, Cu, Cr or Pb or 1% of Zn in its leaves on a dry weight basis is called a hyperaccumulator, irrespective of the metal concentration in the soil [20].

The ability of a metal hyperaccumulating plant could be characterized at four points. First, metal accumulation as a function of soil metal concentrations, physical and chemical properties of soil, physiological state of the plant, *etc*. Second, specificity of metal uptake, transport and accumulation in the biomass. Third, the physiological, biochemical and molecular mechanisms of accumulation and hyperaccumulation. Fourth, the biological and evolutionary significance of metal accumulation. In general, a high level of tolerance to heavy metals could rely on either reduced uptake or increased plant internal sequestration, which is manifested by an interaction between a genotype and its environment [22, 23]. The hyperaccumulator plants possess a range of potential mechanisms involved in detoxification of heavy metals inside plant cells.

i.   Binding to the cell wall.

ii. Reduced uptake or efflux pumping of metals at the plasma membrane.

iii. Chelation of the metal in the cytosol by various ligands, such as phytochelatins, metallothioneins, and metal-binding proteins.

iv. Repair of stress-damaged proteins.

v.   Compartmentalization of metals in the vacuole by tonoplast-located transporters.

Even though, many plants accumulate large amounts of As (≥ 1 μg g^{-1} biomass) [24], they cannot be grouped as hyperaccumulators since As accumulation in these plants occurs very slowly over an extended period of time and a large portion is sequestered into the roots and BF is insignificant.

METAL TOLERANCE AND ACCUMULATION IN PLANTS

Metal tolerance is an evolutionary phenomenon, demonstrated by comparing the growth of plants growing in mining sites with plants outside mining sites grown in non-contaminated soils [25]. Tolerance to heavy metals is generally under major gene control, though, the ability of a species to evolve tolerance seems to depend on the presence of tolerance genes at low frequency in normal populations [26]. Evolutionary studies have shown that one of the mechanisms for metal tolerance is uptake, not exclusion [25]. Hyperaccumulators are the extreme forms of tolerance. Tolerance to metals by plants is governed by the operation of uptake systems that are under direct control of metal concentrations in the soil solutions. Most plant species possess two uptake systems: the highly inducible high-affinity system operational at low concentrations (such as the high affinity phosphate uptake system under low phosphate status) and the constitutive low-affinity system that is effective at high concentrations [26]. In arsenate-tolerant *Holcus lantus* plants [27], the high-affinity system appeared to be lost enabling the plants to take up arsenate more slowly than non-tolerant plants, but the tolerant plants were capable of accumulating As to high concentrations over a long period of time. In non-tolerant plants, both high and low-affinity uptake systems are present and their operation depends upon As concentration. Plants that adapt to high As levels evolved tolerance by suppressing the high affinity phosphate-arsenate uptake system [26].

The studies on As metabolism in terrestrial plant fall into two classes [28]: those that did not characterize the form of As extracted from plant tissues as chelated in the tissues or not and those that did. Arsenate resistance has been identified in a range of plant species including *Holcus lanatus, Calluna vulgaris* and *Silene vulgaris* [29-32]. Despite this understanding of the processes controlling decreased arsenate uptake, arsenate resistant plants can still accumulate considerable levels of As in their tissues, for example, 3470 µg g^{-1} As in *Agrostis tenuis* and 560 µg g^{-1} As in *H. lanatus* [24]. Arsenic-resistant plants either compartmentalize and /or transform As to other less phytotoxic As species in order to withstand high cellular As burdens [30]. Exposure to inorganic As species results in the generation of reactive oxygen species (ROS) [33] and leads to the synthesis of enzymatic antioxidants such as superoxide dismutase (SOD), catalase and glutathione-S-transferase, and nonenzymatic antioxidants, for example glutathione and ascorbate [33-36]. Glutathione is also utilized for the synthesis of phytochelatins (pcs) ([γ-glutamate-cysteine]n-glycine) in plants, upon exposure to inorganic metal ions like As. Following the reduction of arsenate to arsenite in plants, As may potentially be further metabolized to methylated species leading to further oxidative stress [37]. The exception to the rule that plants down regulate their arsenate uptake to tolerate high levels of arsenate is a plant that exhibits As hyperaccumulation. The Brake fern, *Pteris vittata*, has the remarkable ability to hyperaccumulate As in its aerial biomass [7, 38, 39], reaching levels ~100-fold higher in the shoots than soil concentrations [7]. This capability to hyperaccumulate As is, to date, unique. Many plant species were identified as occurring on nickel, copper and serpentine mine sites in Zimbabwe [9]. Interestingly the fern, *Pteris vittaria*, was also identified in one of the gold mines studied. In addition to the remarkable ability of *P. vittata* to tolerate high internal As burdens, it is also capable of extracting low levels of arsenate from soils. Some plant species have evolved mechanisms to mobilize phosphate from soils including the production of proteoid roots or acidification of the rhizosphere [40]. Adaptations in root morphology / rhizosphere chemistry may have deployed by *P. vittata* to mobilize arsenate, although the mechanisms used to mobilize arsenate will also mobilize phosphate. The plant is also able to hyperaccumulate phosphate [7]. This helps to explain its ability to tolerate very high concentrations of arsenate as non-resistant *H. lanatus* can become arsenate resistant under high phosphorous status [30].

STATE-OF-ART UTILIZATION OF FERNS FOR METAL REMEDIATION

Phytoremediation of As by Chinese-Brake

Although ferns belong to the evolutionarily highest group of vascular plants, they are not important crops and have little impact on the humans [41]. This paradigm may change upon ascertaining the phytoremediation potential of Chinese Brake fern by exploring the physiological and molecular mechanisms of As hyperaccumulation. For decontamination of As-polluted sites using Chinese-Brake fern, phytoextraction or phytomining is the most suitable mechanism. This is also the mechanism which is generating interest in the scientific and technological world due to its ease in implementation. In the past, attempts to phytoremediate contaminated sites by employing phytoextraction were unsuccessful. This is because the total amount of metal removed from a site is a product of metal concentration in the harvested plant material and the total harvested biomass [42-44]. Most of the known hyperaccumulators grow very slowly and produce a low biomass. On the contrary, the Chinese-Brake fern is equipped with all the properties required by an ideal hyperaccumulating plant portended for phytoremedaition purposes. Chinese Brake fern has emerged as a prospective system for As phytoremediation due to its appropriateness.

Large Biomass

Ma *et al.* [7] discovered that As hyperaccumulation by the Chinese Brake fern was accompanied by the development of a large biomass. For instance, the Chinese Brake fern grown in As- contaminated soils produced a total dry biomass of 18 g plant^{-1} after 18 weeks of growth [45]. This is much faster than the growth rates of other hyperaccumulators such as a well-known Zd, Cd hyperaccumulator, *T. Caerulescens* [46]. Arsenic concentrations in the aerial biomass were as high as 2.3% [7] and most of the As taken up from soil was translocated to the fronds (90%). Very little As was retained in the roots. Furthermore, As concentration was increased with frond age [45] indicating that As is mobile in the plant. About 26% of the original soil As was removed by the fern after 20 weeks of growth [45].

Treatment of As-Contaminated Groundwater with Chinese-Brake Fern

Usefulness of Chinese Brake fern to remove As from contaminated groundwater was assessed [47]. It was demonstrated that Chinese-Brake fern effectively reduced the arsenic concentration from 46 to less than 10 µg L^{-1} in 3 days. In a repeated treatment of these plants, they continued to take up arsenic from the groundwater, albeit at a slower rate (from 46 to 20 µg L^{-1} during the same time). Young fern plants were more efficient in removing arsenic than older fern plants of similar size. The addition of a supplement of phosphate-free Hoagland solution to the groundwater had little effect on arsenic removal, but the addition of phosphate significantly reduced its arsenic affinity and, thus, inhibited arsenic removal. This study suggested that Chinese Brake fern has the potential to remove arsenic from groundwater.

LIMITATION OF FERNS FOR PHYTOREMEDIATION APPLICATION

- ➜ Low biomass: To date, there are more than 400 plant species known [21] to hyperaccumulate heavy metals but most of them fall short of biomass. Ferns are slow growing and have shallow roots that do not extend to a great depth. Therefore, the efficiency of metal extraction by ferns is limited to the zone of root penetration and remediation of soil substrate lower down is not practicable.

➔ Requirement of optimum agro-climatic conditions: The ferns being water loving species endemic to sub-tropical and humid geographical regions their phytoremediation efficiency will be limited to damp soils. Phytoremediation of dry geographical regions seems not practicable.

➔ Constraints for field application: The aerial parts of ferns are feeble and prone to mechanical sear, which is likely to be encountered under field conditions due to anthropogenic activities and natural climatic changes. Therefore, the phytoremediation application sites will require specific design to provide mechanical support to the plants to maintain their physiology for proper functioning. Depending on the sites and region designing appropriate systems will be required for *in situ* phytoremediation application using ferns.

➔ Post harvest management: Management of the plant biomass with high concentrations of toxic metal ions is essential. The biomass should not decompose on the application site otherwise the metals will re-enter to the soil. Furthermore, management practices for easy harvest of plant biomass and a method for its safe disposal has to be developed. Handling of the biomass with toxic metals and a method for safe disposal will depend on the type of metal. Depending on the efficacy of the method, the plant biomass devoid of the metals can be utilized as manure. However, further investigations into determining the most appropriate agronomic practices are also needed to enhance plant growth and arsenic uptake in order to obtain maximum soil arsenic removal using ferns.

RESEARCH AND DEVELOPMENT STRATEGIES FOR PHYTOREMEDIATION TECHNOLOGY DEVELOPMENT

The R&D areas prospective to enhance metal hyperaccumulation can be projected under functionalities of the different plant components [38]. The rhizospheric zone, stelar zone and the aerial parts are the three principal components of the plant, which need to be targeted to enhance metal hyperaccumulations for phytoremediation. The strategies and scope to enhance their function for phytoremediation through biotechnological interventions are highlighted below.

1. Rhizospheric zone: root and soil system, which is the most dynamic component.

2. Zone for transportation of the element: the vascular conductive system.

3. Aerial biomass: the leaves ultimate sink site.

❖ Rhizospheric

➜ Ameliorate rhizospheric ambient environment conducive for availability of element ions

➜ Enhance element uptake by root cells

➜ Modulate to promote growth activity of the plant

➜ Toxin tolerance at the root zone

❖ Conductive zone for ascent of sap

➜ Transformation to most mobile phase

➜ Flow through vascular system

➜ Chaperons to vehicle the ions

➜ Quick transportation to the activity site

❖ Aerial biomass / sink site

➜ Physiological adaptations to segregate toxicants from essentials

➜ Activate biochemical mechanism(s) to initiate reaction

➜ Chelate/reduce/convert/amalgate

➜ Transform to best managed element species

➜ Efflux from cellular interiors

➜ Make the physical site ready for storage

➜ Compartmentalize and sustain

➜ Enhance hyperaccumulation

➜ Sequestration in the vacuoles

➜ Alteration of metabolic pathway

➜ Oxidative stress prevention and enhance production of intracellular chaperons

➜ Activation of differential defensive response and other adaptive mechanisms

➜ Enhance biomass production

PROGRESS IN ENGINEERING HYPERACCUMULATION IN PLANT SYSTEMS

Out of the several approaches, the following two could be prospective due to practical feasibility.

1. Overexpression of metal transporter genes leading to enhanced metal uptake, translocation and/or sequestration depends on the tissues where the gene is expressed (root, shoot, vascular tissue, or all) and on the intracellular targeting (*e.g.,* cell membrane, vacuolar membrane).

2. The overexpression of genes involved in synthesis of metal chelators leading to enhanced metal uptake, as well as enhanced metal translocation and/or sequestration, depending on the type of chelator and its location.

The initial step in As metabolism is the enzymatic reduction of As(V) to As(III) mediated by the enzyme arsenate reductase (ACR) [48, 49] and thereby As(III) amenable for conversion to a non-toxic form for efflux or chelation. Reduced activity of this enzyme in the roots facilitates As uptake and enhance accumulation in the above-ground biomass. Using this approach, genetically engineered transgenic *Arabidopsis* plant that transported As to the above-ground biomass, reduced it to As(III) and sequestered As(III) as thiol-peptide complexes through co-expression of *E. coli* arsenate reductase (*ars*C) and γ-ECS gene was developed [50]. The genetically engineered *A. thaliana* plants with reduced activity of arsenate reductase enzyme in the roots enhanced As accumulation (10-16-fold more) in the above-ground biomass for phytoremediation. In another approach, overexpression of phytochelatin synthase (PCS) gene [51] in the *Arabidopsis* plant system increased tolerance to As and Hg. Overexpression of *acds* gene in transgenic canola (*B. napus*) plants conferred As resistance and the seeds germinated up to 150 mg L^{-1} As(V) [52].

GENES WITH POTENTIAL FOR PHYTOREMEDIATION

The knowledge base of the genetics and biochemical processes involved in metal uptake, transport and storage, and a greater insight into the process of hyperaccumulation, is essential for the development of plants with improved phytoremediation capability [53-59]. Physiological and classical genetic studies have been complemented by molecular studies, in particular transcriptome analysis. Presently, it is not easy to reconcile the results of the different research approaches applied. There is a specific molecular mechanism for uptake, transport and sequestration of each metal. Extensive progress has been made in identifying genes and proteins involved in metal uptake by plants [59-61]. Hyperaccumulators are a good source of genes suitable for phytoremediation [59, 62, 63]. The regulatory control and use of tissue specific promoters offer great promise to develop plants for removal of elemental pollutants and radionuclides. Hyperaccumulators are loaded with acids and acid anions that have a function in metal storage or plant internal metal transport [59, 64]. Transgenic plants could be developed to secrete metal-selective ligands into the rhizosphere that could specifically solubilize elements for phytoremediation [57]. Finding simple molecules with selective chelation ability, which plants can make and secrete into the rhizosphere and simultaneously engineering plants with capability for transporting protein for the metal-chelates could be an area for research and development. In Ni hyperaccumulator, free histidine in xylem exudates was found as a metal chelator [65]. Histidine concentrations in the xylem exudates can be modified for increasing Ni accumulating capacity in plant. Other potential mediators of metal sequestration and accumulation include cation diffusion facilitation family (CDF) [66]. Cellular targeting, especially in the vacuoles, is important since the heavy metals can be kept in a safe compartment without disturbing the cellular functions. Engineering vacuolar transporters, preferably in specific cell types, is a second-generation approach for phytoremediation [59]. Another alternative is to create artificial metal sinks in the shoot by enhancing metal binding sites. Great strides have been made in the development of transgenic plants for phytoremediation, but majority of genes has been transferred from other organisms to plants [67-75]. Biomass of known hyperaccumulators can be altered by introduction of genes that affect phytohormone synthesis resulting in enhanced biomass. The biosynthetic pathways for most of the plant hormones have been elucidated and genes encoding key enzymes have been cloned [76].

At least three different engineering approaches can be envisioned to enhance metal uptake, which include; (i) enhancing the number of uptake sites, (ii) alteration of specificity of uptake system to reduce competition by unwanted cations, and (iii) increasing intracellular binding and sequestration. Some of the different genes, which have been used for the development of transgenic plants and those having direct or indirect relevance in enhancing metal accumulation have been summarized [77]. These genes have potential roles in modulating the physiological and biochemical processes of plants and could be of use for development of transgenic hyperaccumlator plants. They mediate direct and indirect functions in metal accumulation and tolerance and forthcoming for genetic engineering to enhance metal tolerance in hyperaccumulator and nonhyperaccumulator plants.

FUTURE PROSPECTS

Phytoremediation is still in its infancy, but it is being used in some in-field remediation [78, 79]. However, there are several issues. One of which is cost. It is widely claimed that phytoremediation is a much more economical remediation technology compared to most other remediation techniques. However, the costs can vary widely depending on the site factors and the plant(s) being used to remediate the sites. Another issue is time. The amount of time needed to fully remediate a site is, again, very dependent on the plant and site characteristics [78]. It has been shown in multiple studies that trace element metabolisms in plants are genetically regulated and can be manipulated, leading to plants with altered metal tolerance, accumulation and/or capacity for biotransformation. When natural plant processes were accelerated by genetic engineering, 2 to 3 fold increase in metal accumulation in the transgenic plants was reported [75]. This would potentially reduce the cost of phytoremediation to same extent, if the same results hold true in the field. Furthermore, the introduction of a new pathway has led to plants that can detoxify (in case of Hg, As and Se) in ways that other plants cannot. This is potentially valuable. As discovery of more metal resistance-related genes is facilitated by genome sequencing, many new possibilities are opening up for the creation of new transgenics with favorable properties for phytoremediation. Physiological studies in *T. Caerulescens* have revealed considerable diversity among populations with regard to the capacities and the specific metal-affinity patterns [80]. Microarray analyses have advanced the knowledge of hyperaccumulation by providing promising candidate genes [81]. Identification of the regulatory genes enables transgenic strategies to engineer plants with higher tolerance capacities or modified accumulation of trace metals. The promising strategies are summarized: (1) The many newly discovered metal transporters, including the ones from hyperaccumulator plants (ZNT1, (tgmtp1: T. Goesingense CDF members) may be overexpressed in high biomass plant species, targeted to different tissues and intracellular locations [59]. (2) Nicotianamine overproduction may be an avenue to manipulate metal translocation and tolerance, with special reference to iron uptake, NA being the precursor of phytosiderophores [82]. Overproduction of NA is feasible *via* overexpression of enzymes from the NA biosynthesis pathway, for which genes have been cloned [62, 83]. (3) Overexpression of phytochelatin synthase (PCS) mediating PC synthesis from GSH may further enhance metal tolerance and accumulation [56, 84-86]. The overexpression of the vacuolar transporter responsible for shuttling the PC-metal complex into the vacuole also enhance metal tolerance and accumulation [75, 86]. (4) Overproduction of histidine, which confers Ni tolerance [65, 87, 88] can be achieved and the genes involved in His biosynthesis have been cloned [89]. (5) Another area is molecular biology of the rhizosphere. Manipulation of the quality and quantity of root-released compounds [90, 91] offer a promising alternative strategy to affect metal uptake or exclusion. Nevertheless, winning phytoremediation strategies have involved the introduction into plants of microbial detoxification pathways for elements such as Hg [92] and As [30, 52, 79].

CONCLUSIONS

The development of phytoremediation technologies is continuing, involving transgenic and non-transgenic approaches [93]. Several difficulties are encountered when attempting to transfer emerging technologies from the laboratory into the field. Legislative hurdles and regulatory uncertainties often limit access to contaminated experimental field sites and collect relevant information. Moreover, it has gradually become clear that the lack of a comprehensive understanding of the complex and tightly regulated metal homeostatic network in plants is still a major bottleneck in the development of phytoremediation technologies [59]. With the engineering of a high-biomass, multi-metal tolerant, metal-accumulating phytoremediator plant in mind, the modification of expression levels and/or introduction of multiple genes appear to be as daunting as it is indispensable. This insight may have triggered the search for hyperaccumulator genes with metalloregulatory functions. It is necessary to be able to shed light on some of the poorly understood phenomena observed in the field, such as the positive response of hyperaccumulator root growth to high, localized metal concentrations in the soil or the role of the interaction of microorganisms with plant roots in modulating bioavailability of metals. Whatever the prospects are of finding a 'master regulator', answering the questions as to how metals are sensed by higher plants, and how a signal is generated and transduced to result in a response to altered metal availability, will provide challenging and fascinating avenues for research in the years to come. Plant biotechnologists have not yet used genes from ferns (and other lower plants) for improving angiosperm crops. Together, these new developments will likely give rise to much new information about metal metabolism in plants in the near future and may lead to the fruitful applications in environmental cleanup.

ACKNOWLEDGEMENTS

This work has been carried out under the aegis of Council of Scientific and Industrial Research (CSIR), Govt. of India at the National Environmental Engineering Research Institute (NEERI), Nagpur. The authors are grateful to the Director NEERI for his guidance in preparation of the manuscript.

CONFLICT OF INTEREST

The author(s) confirm that this chapter content has no conflict of interest.

REFERENCES

[1]   Sanchez-Baracaldo P. Phylogenetic and biogeography of the neotropical fern genera Jamesonia and Eriosorus (Pteridaceae). Am J Bot 2004; 91: 274-84.

[2]   Porembski S, Barthlott W. Granitic and gneissic outcrops (inselbergs) as centers of diversity for desiccation-tolerant vascular plants. Plant Ecol 2000; 151: 19-28.

[3]   Medina E, Cuevas E, Popp M, Lugo AE. Soil salinity, sun exposure, and growth of *Acrostichum aureum* the mangrove fern. Bot Gazette 1990; 151: 41-49.

[4]   Francis PC, Petersen RL. Effect of copper, cadmium, and zinc on percent spore germination of the cinnamon fern (Osmunda Cinnamomea) and the sensitive fern (*Onoclea sensibilis*). Bullet Environ Contaminat Toxicol 1983; 30: 559-66.

[5]   Nishizono H, Suzuki S, Ishii F. Accumulation of heavy metals in The metal-tolerant fern *Athyrium yokoscense*, growing on various environments. Plant Soil 1987; 102: 65-70.

[6]   Kamachi H, Komori I, Tamura H, *et al.* Lead tolerance and accumulation in the gametophytes of the Fern *Athyrium yokoscense*. J Plant Res 2005; 118: 137-45.

[7]   Ma LQ, Komar KM, Tu C, Zhang W, Cai Y, Kennelley ED. A fern that hyperaccumulates arsenic. Nature 2001; 409: 579.

[8]   Vogt T. Geochemical and geobotanical ore prospecting. Some notes on the vegetation at the ore deposits at Romp. K Norske Vidensk Selsk Skr 1942; 15: 21-4.

[9]   Wild H. Geobotanical anomalies in Rhodesia. I. The vegetation of copper-bearing soils. Kirkia 7. Exp Bot 1968; 40: 119-28.

[10]   Page CN. Ferns: Their Habitats in British and Irish Landscape. New Naturalist Series, Collins, London 1998.

[11]   Kruckeberag R. Ferns associated with ultramafic rocks in the Pacific Northwest. Amer Fern J 1964; 54: 113-26.

[12] Sela M, Garty J, Ellor A. The accumulation and the effect of heavy metals on the water fern, *Azollafiliculoides.* New Phytol 1989; 112: 7-12.

[13] Sen AK, Mondal NG. Removal and uptake of copper by salvinia natans from waste water. Water Air Soil Pollut 1990; 49: 1-16.

[14] Gupta M, Devi S. Uptake and toxicity of cadmium in aquatic ferns. J Environ Biol 1995; 16: 131-6.

[15] Necholas PB, Couch JD, AL-Hamdani SH. Selected physiological responses of *Salvinia minima* to different chromium concentrations. Aquatic Bot 68: 313-9.

[16] Ozaki T, Enomoto S, Minm Y, Ambe S, Makjde Y. A survey of trace elements in pteridophytes. Biol Trace Elem Res 2000; 74: 259-73.

[17] Meharg AA. Variation in arsenic accumulation-hyperaccumulation in ferns and their allies. New Phytol 2003; 157: 25-31.

[18] Sridokchan W, Markich S, Visoottiviseth P. Arsenic tolerance, Accumulation and Elemental Distribution in Twelve Ferns: A Screening study. Aust J Ecotoxicol 2005; 11: 101-10.

[19] Francesconi K, Visoottiviseth P, Sridokchan W, Goesslero W. (Arsenic species in an arsenic hyperaccumulating fern, *Pityrogramma calomelanos*: a potential phytoremediator of arsenic-contaminated soils. Sci Total Environ 2002; 284: 27-35.

[20] Baker AJM, Walker PL. Ecophysiology of Metal Uptake by Tolerant Plants. In: Shaw AJ, Ed. Heavy Metal Tolerance in Plants: Evolutionary Aspects. Florida, Boca Raton, CRC Press, 1990; pp. 155-177.

[21] Brooks RR. Phytochemistry of Hyperaccumulator. In: Brooks RR, Ed. Plants that Hyperaccumulate Heavy Metals - Their Role in Phytoremediation, Microbiology, Archeology, Mineral Exploration and Phytomining. New York, CAB International, 1998; pp. 15-54.

[22] Macnair MR, Rlston GH, Smith SE. The genetics of metal tolerance and accumulation in higher plants. Pp. 235-248. In: Terry N, Banueoles G, Eds. Phytoremediation of Contaminated Soil and Water. London, Lewis Publishers, 2000; pp.235-48.

[23] Hall AH. Chronic arsenic poisoning. Toxicol Lett 2002; 128: 69-72.

[24] Porter EK, Peterson PJ. Arsenic accumulation by plants on mine wastes (United Kingdom). Sci Total Environ 1975; 4: 365-71.

[25] Baker AJM, Brooks R Reeves R. Growing for gold and copper and zinc. New Scient 1988; 117: 44-8.

[26] Macnair MR. The Evolution of Plants in Metal-Contaminated Environments. In: Bijlsma R, Loeschcke V, Eds. Environmental Stress, Adaptation and Evolution. Boston, Birkhauser Verlag, 1997; pp. 3-24.

[27] Mehrag AA, Macnair MR. An altered phosphate uptake system in arsenate-tolerant *Holcus lantus* L. New Phytol 1990; 116: 29-35.

[28] Meharg AA, Hartley-Whitaker J. Arsenic uptake and metabolism in arsenic resistant and nonresistant plant species. New Phytol 2002; 154: 29- 43.

[29] Paliouris G, Hutchinson TC. Arsenic, cobalt, and nickel tolerances in two populations of *Silene vulgaris* (Moench) Garcke from Ontario, Canada. New Phytol 1991; 117: 449-59.

[30]   Mehrag AA. Integrated tolerance mechanisms: constitutive and adaptive plant responses to elevated metal concentrations in the environment. Plant Cell Environ 1994; 17: 989-93.

[31]   Fitter AH, Wright WJ, Williamson L, Belshaw M, Fairclough J, Meharg AA. The Phosphorus Nutrition of Wild Plants and the Paradox of Arsenate Tolerance: Does Leaf Phosphate Concentration Control Flowering? In: Lynch JP, Deikman J, Eds. Phosphorus in Plant Biology: Regulatory Roles in Molecular, Cellular, Organism and Ecosystem Processes. Am Soc Plant Physiol 1998; 39-51.

[32]   Sharples JM, Meharg AA, Chambers SM, Cairney JWG. Symbiotic solution to arsenic contamination. Nature 2000; 404: 951-2.

[33]   Hartley-Whitagerk A, Insworth VR, Ten Bookum W, Schat H, Mehrag AA. Phytochelatins are involved in differential arsenate tolerance in *Holcus lantus*. Plant Physiol 2001; 126: 299-306.

[34]   Alscher RG. Biosynthesis and antioxidant function of glutathione in plants. Physiol Plant 1989; 77: 457-64.

[35]   Mylona PV, Polidoros AN, Scandalios JG. Modulation of antioxidant responses by arsenic in maize. Free Radical Biol Med 1998; 25: 576-85.

[36]   Dat J, Vandenabeele S, Vranova E, Van Montagu M, Inze D, Van Breusegem F. Dual action of the active oxygen species during plant stress responses. Cellular Molecul Life Sci 2000; 57: 779-95.

[37]   Zaman K, Pardini RS. An overview of the relationship between oxidative stress and mercury and arsenic. Toxic Substance Mech 1996; 15: 151-81.

[38]   Sarangi BK, Kalve SK, Pandey RA, Chakrabarti T. Transgenic plants for phytoremediation of arsenic and chromium to enhance tolerance and hyperaccumulation, Transgenic Plant J 2009; 3: 57-86.

[39]   Kalve S, Sarangi BK, Pandey RA, Chakrabarti T. Arsenic and chromium hyperaccumulation by an ecotype of *Pteris vittata*-prospective for phytoextraction from contaminated water and soil. Curr Sci 2011; 100: 888-94.

[40]   Marschner H. Mineral Nutrition of Higher Plants. Academic Press, London, 1995.

[41]   Raven PH, Evert RF, Eichhorn SE. Seedless Vascular Plants. In: Anderson S, Matalski E, Eds. Biology of Plants. New York, Worth Publishers, 1992; pp.323-324.

[42]   Baker AJM, Reeves RD, McGrath SP. *In Situ* Decontamination of Heavy Metal Polluted Soils Using Crops of Metal-Accumulating Plants-A Feasibility Study. In: Hinchee RE, Olfenbuttel RF, Eds. *In Situ* Bioreclamation. Stoneham, MA Buttenvorth- Heinemann, 1991; pp. 539-544.

[43]   Salt DE, Smith RD, Raskin I. Phytoremediation. Annu Rev Plant Physiol Plant Mol.Biol 1998; 49: 643-68.

[44]   Glass DJ. U. S. and International Markets for Phytoremediation, 1999-2000. D. Glass Associates, Inc., Needham, MA, USA 1999.

[45]   Tu C, Ma LQ. Effects of arsenate and phosphate on their accumulation by an arsenic hyper accumulator *Pteris vittata* L. Biomed Life Sci 2003; 249: 373-82.

[46]   Lombi E, Zhao FJ, Dunham SJ, McGrath SP. Cadmium accumulation in populations of *Thlaspi caerulescens* and *Thlaspi goesingense*. New Phytol 2000; 145: 11-20.

[47] Tu S, Ma LQ, Fayig AO, Zillioux EJ. Phytoremediation of arsenic-contaminated groundwater by the arsenic hyperaccumulating fern *Pteris vittata* L. Int J Phytoremediat 2004; 6: 35-47.

[48] Rosen BP. Families of arsenic transporters. Trends Microbiol 1999; 7: 207-12.

[49] Rosen BP. Transport and detoxification systems for transition metals, heavy metals andmetalloids in eukaryotic and prokaryotic microbes. Comp Biochem Physiol A 2002; 133: 689-93.

[50] Dhankher OP, Li Y, Rosen BP, *et al.* Engineering tolerance and hyperaccumulation of arsenic in plants by combining arsenate reductase and gamma-glutamylcysteine synthetase expression. Nature Biotechnol 2002; 20: 1140-5.

[51] Li Y, Dhankher OP, Carreira L, *et al.* Overexpression of phytochelatin synthase in *Arabidopsis* leads to enhanced arsenic tolerance and cadmium sensitivity. Plant Cell Physiol 2004; 45: 1787-97.

[52] Nie L, Shah S, Rashid A, Burd GI, Dixon GD, Glick BR. Phytoremediation of arsenate contaminated soil by transgenic canola and the plant growth-promoting bacterium *Enterobacter cloacae* CAL2. Plant Physiol Biochem 2002; 40: 355-61.

[53] Baker AJM, McGrath SP, Reeve RD, Smith JAC. Metal Hyperaccumulator Plants: A Review of the Ecology and Physiology of a Biological Resource for Phytoremediation of Metal Polluted Soils. In: Terry N, Banuelos G, Eds. Phytoremediation of contaminated soil and water. Florida, Boca Raton, CRC Press, 2000; pp. 85-107.

[54] Karenlampi S, Schat H, Vangronsveld J, *et al.* Genetic engineering in the improvement of plants for Phytoremediation of metal polluted soils. Environ Pollut 2000; 107: 225-31.

[55] Meagher RB. Phytoremediation of toxic elemental and organic pollutants. Curr Opinion Plant Biol 2000; 3: 153-62.

[56] Clemens EJ, Kim ND, Schroeder JI. Tolerance to toxic metals by a gene family of phytochelatin synthases from plants and yeast. EMBO J 1999; 18: 3325-33.

[57] Pilon-Smits E, Pilon M. Phytoremediation of metals using transgenic plants. Crit Rev Plant Sci 2002; 21: 439-56.

[58] Pollard AJ, Powell KD, Harper FA, Smith JAC. The genetic basis of metal hyper accumulation in plants. Crit Rev Plant Sci 2002; 21: 539-66.

[59] Verbruggen N, Hermans C, Schat H. Molecular mechanisms of metal hyperaccumulation in plants. New Phytol 2009; 181:759-76.

[60] Eide D, Broderius M Fett J, Guerinot ML. A novel iron- regulated metal transporter from plants identified by functional expression in yeast. PNAS USA 1996; 93: 5624-8.

[61] Guerinot ML. Improving rice yields — ironing out the details. Nature Biotechnol 2001; 19: 417-8.

[62] van de Mortel JE, Villanueva LA, Schat H, *et al.* Large expression differences in genes for iron and zinc homeostasis, stress response, and lignin biosynthesis distinguish roots of *Arabidopsis thaliana* and the related metal hyperaccumulator *Thlaspi caerulescens*. Plant Physiol 2006; 142: 1127-47

[63] Van de Mortel JE, Schat H, Moerland PD, *et al.* Expression differences for genes involved in lignin, glutathione and sulphate metabolism in response to cadmium in *Arabidopsis*

thaliana and the related Zn/Cd-hyperaccumulator *Thlaspi caerulescens*. Plant, Cell Environ 2008; 31: 301-24.

[64]   Callahan D, Baker A, Kolev S, Wedd. A. Metal ion ligands in hyperaccumulating plants. J Biol Inorgan Chem 2006; 11: 2-12.

[65]   Kramer U, Talke IN, Hanikenne M. Transition metal transport. FEBS Lett 2007; 581: 2263-72.

[66]   Peiter E, Montanini B, Gobert A, *et al*. A secretory pathway-localized cation diffusion facilitator confers plant manganese tolerance. PNAS USA 2007;104: 8532-7.

[67]   Rugh CL, Wilde HD, Stack NM, Thompson DM, Summers AO, Meagher R.B. Mercuric ion reduction and resistance in transgenic *Arabidopsis thaliana* plants expressing a modified bacterial merA gene. PNAS USA 1996; 93: 3182-7.

[68]   Rugh CL, Bizily SP, Meagher RB. Phytoreduction of Environmental Mercury Pollution. In: Phytoremediation of Toxic Metals — Using Plants to Clean up the Environment. Raskin I, Ensley BD, Eds. New York, Wiley, 2000; pp. 151-71.

[69]   Grichko VP, Filby B, Glick BR. Increased ability of transgenic plants expressing the bacterial enzyme ACC deaminase to accumulate Cd, Co, Cu, Ni, Pb, and Zn. J Biotechnol 2000; 81: 45-53.

[70]   Lin ZQ, Schemenauer RS, Cervinka V, Zayed A, Lee A, Terry N. Selenium volatilization from the soil - *Salicornia bigelovii* Torr. treatment system for the remediation of contaminated water and soil in the San Joaquin Valley. J Environ Qual 2000; 29: 1048-56.

[71]   Pilon-Smits EAH, Zhu YL, Sears T, Terry N. Overexpression of glutathione reductase in *Brassica juncea*: effects on cadmium accumulation and tolerance. Physiol Plant 2000; 110: 455-60.

[72]   Harada E, Choi YE, Tsuchisaka A, Obata H, Sano H. Transgenic tobacco plants expressing a rice cysteine synthase gene are tolerant to toxic levels of cadmium. J Plant Physiol 2001; 158: 655-61.

[73]   Barcelo J, Poschenrieder C, Tolrà RP. Importance of Phenolics in Rhizosphere and Roots for Plant Metal Relationships. In: Gobran G, Ed. Extended Abstracts 7th ICOBTE, Upsala, 2003; pp. 162-163.

[74]   Gisbert C, Ros R, de Haro A, *et al*. A plant genetically modified that accumulates Pb is especially promising for Phytoremediation. Biochem Biophys Res Commun 2003; 303:440-5.

[75]   Dhankher OP, Rosen BP, McKinney EC Meagher RB. Hyperaccumulation of arsenic in the shoots of Arabidopsis silenced for arsenate reductase, ACR2. PNAS USA 2006; 103: 5413-5.

[76]   Woodward, A, Bartel B. Auxin: regulation, action and interaction. Ann Bot 2005; 95: 707-35.

[77]   Sarangi BK, Dash T, Pandey RA. Engineering phytoremediation potentiality of plants through hyperaccumulation in plant biomass - with reference to Arsenic and Chromium. J Plant Sci Res 2010; 26: 113-45.

[78]   Schnoor JL. Phytoremediation of soil and groundwater. GWRTAC TechnoloEvaluation Report TE-02-01, 2002.

[79] USEPA (United States Environmental Protection Agency) (2002) Arsenic treatment technologies for soil, waste and water. Report EPA-542-R-02-004. Washington. Aust J Ecotoxicol 2002; 11: 101-10.

[80] Assunçao AGL, Bleeker P, ten Bookum WM, Vooijs R, Schat H. Intraspecific variation of metal preference patterns for hyperaccumulation in *Thlaspi caerulescens*: evidence from binary metal exposures. Plant Soil 2008; 303: 289-99.

[81] Chakrabarti S, Sen R. Biotechnology - applications to environmental remediation in resource exploitation. Curr Sci 2009;97: 768-75.

[82] Higuchi K, Suzuki K, Nakanishi H, Yamaguchi H, Nishizawa NK, Mori S. Cloning of nicotianamine synthase genes, novel genes involved in the biosynthesis of phytosiderophores. Plant Physiol 1999; 119: 471-9.

[83] Takahashi M, Yamaguchi H, Nakanishi H, Shioiri T, Nishizawa NK, Mori S. Cloning two genes for nicotianamine aminotransferase, a critical enzyme in iron acquisition (strategy II) in graminaceous plants. Plant Physiol 1999; 121: 947-56.

[84] Ha S-B, Smith AP, Howden R, *et al.* Phytochelatin synthase genes from *Arabidopsis* and the yeast *Schizosaccharomyces pombe*. Plant Cell 1999; 11: 1153- 63.

[85] Vatamaniuk OK, Mari S, Lu YP, Rea PA. AtPCS1, a phytochelatin synthase from *Arabidopsis*: isolation and *in vitro* reconstitution. PNAS USA 1999; 96: 7110-5.

[86] Li Y, Dankher OP, Carreina L, Smith A, Meagher R. The shoot-specific expression of γ-glutamylcysteine synthetase directs the long-distance transport of thiol-peptides to roots conferring tolerance to mercury and arsenic. Plant Physiol 2006; 141: 288-98.

[87] Kramer U, Chardonnens AN. The use of transgenic plants in the bioremediation of soils contaminated with trace elements. Applied Microbiol Biotechnol 2001; 55:661-72.

[88] Ingle RA, Smith JAC, Sweetlove LJ. Responses to nickel in the proteome of the hyperaccumulator plant *Alyssum lesbiacum*. BioMetals 2005; 18: 627-41.

[89] Persans, M.W., Yan, X.G., Patnoe, Jean-M., Kramer,U. and Salt, D.E. Molecular dissection of the role of histidine in nickel hyperaccumulation in *Thlaspi goesingense* (Halacsy). Plant Physiol 1999; 121: 1117-26.

[90] Gleba D, Borisjuk NV, Borisjuk LG, *et al.* Use of plant roots for phytoremediation and molecular farming. PNAS USA 1999; 96: 5973-7.

[91] Dusenkov S, Skarzhinskaya M, Glimelius K, Gleba D, Raskin I. Bioengineering of a Phytoremediation Plant by Means of Somatic Hybridization. Int J Phytoremediat 2002; 4: 117-26.

[92] Bizily SJ, Rugh CL, Meagher RB. Phytodetoxification of hazardous organomercurials by genetically engineered plants. Nature Biotechnol 2000; 18: 213-7.

[93] Kraemer U. Phytoremediation to phytochelatin - plant trace metal homeostasis. New Phytol 2003; 158: 1-9.

Send Orders for Reprints to reprints@benthamscience.net

CHAPTER 5

Thlaspi caerulescens and/or Related Species: Progress and Future Prospects

Katarina Vogel-Mikuš[*]

Biotechnical Faculty, Department of Biology, Večna pot 111, SI-1000 Ljubljana, Slovenia

Abstract: In order to be able to use phytoremediation practices successfully it is necessary to gain knowledge on the behaviour and fate of the metals in soil-plant systems. Metal hyperaccumulator plants such as *Thlaspi caerulescens* of the Brassicaceae family represent an excellent model to study physiological and molecular mechanisms of metal uptake, transport, accumulation and tolerance due to their physiological, morphological and genetic characteristics, and their close relationship to *Arabidopsis thaliana*, the general plant reference species. In this chapter, the progress that has been made in elucidating molecular and physiological mechanisms of metal hyperaccumulation in *Thlaspi caerulescens* (a model Zn, Cd and Ni hyperaccumulator) and its relative *T. praecox* (a Cd and Zn hyperaccumulator) will be reviewed briefly. A special emphasis will be placed on hyperaccumulation of cadmium and interactions of the *Thlaspi spp.* with symbiotic arbuscular mycorrhizal fungi as these topics still need to be more intensively explored.

Keywords: *Arabidopsis halleri*, Brassicaceae, citrate, Cys-rich proteins, *Glomus* species, glucosinolates, glucotropaeolin, hypertolerance, malate, metal transporters, metallophytes, metallotioneins, nicotianamine, organic acids, phytomining, ATP-ase, *Thlaspi caerulescens*, *Thlaspi praecox*, vacuolar sequestration, ZIP family.

INTRODUCTION

The beginning of the 21st century is witnessing irreversible technological and economic development, which, on the other hand, is accompanied by evident over-exploitation of natural resources and acute water and soil pollution. Soil pollution with highly toxic and non-degradable metals represents a major environmental problem, since it threatens the growth of vegetation, and endangers

Address correspondence to Katarina Vogel-Mikus: Biotechnical Faculty, Department of Biology, Večna pot 111, SI-1000 Ljubljana, Slovenia; Tel: + 386 1 3283 416; Fax: + 386 1 2573390; E-mail: katarina.vogel@bf.uni-lj.si

animal and human health [1]. In the last twenty years, however, a significant development has been achieved in the field of developing different technologies for remediation of degraded and metal-polluted areas, ranging from conventional remediation technologies that usually result in total destruction of the physical, chemical and biological properties of soils, to phytoremediation representing a natural, environmentally friendly technique that utilizes suitable plants and their associated micro-organisms for restoring polluted environments [1-3]. The primary motivation behind the development of phytoremediation practices was the potential to develop low cost remediation methods [2]. Although the term, phytoremediation is relatively new, it is actually an age old practice. Research using semi-aquatic plants for treating radionuclide-contaminated waters existed in Russia already at the dawn of the nuclear era [4, 5].

In order to be able to use phytoremediation practices successfully, it is necessary to gain knowledge on the behaviour and faith of metals in soil-plant systems. Therefore research has been oriented to depict the mechanisms involved in metal uptake, transport, accumulation and tolerance, mostly in model, so called "metal hyperaccumulator" plant species. Metal hyperaccumulation is actually one of the metal tolerance strategies that enables survival of the plants in highly metal enriched/ contaminated environments [6]. In general, in metal hyperaccumulator plants enhanced root metal uptake is accompanied by successful metal loading in the xylem and transport to the shoots or leaves. In these aerial plant parts the metals accumulate in extremely high amounts that are highly toxic for other plants and animals, but do not harm the metal hyperaccumulator plants themselves [7, 8]. In fact the term "metal hyperaccumulator plant" was first used to describe plants with an extreme metal accumulation capacity [9], containing more than 1,000 mg kg^{-1} (0,1%) of Ni in dry leaves, which is for an order of magnitude higher than in »normal« plants [10]. An extensive analytical survey on many plant species has been performed in the past decades, resulting in a constantly increasing list of known metal hyperaccumulator species. Approximately 420 species have now been reported, mainly occurring on metal-rich soils in both tropical and temperate zones and belonging to a wide range of unrelated families. Besides Ni, certain plants hyperaccumulate metal(loid)s like As, Se, Cd, Co, Cu, Mn, Zn, Pb and also some others more exotic elements like Sn, Ce, La, Nd, Pr and Tl [3, 11].

Although metal hyperaccumulation is an extremely rare phenomenon in the plant world, exhibited by less than 0.2% of angiosperms [12], metal hyperaccumulator plant species have drawn attention of the scientists due to several reasons. Besides their potential for the use in phytoextraction of heavy metals from polluted soils and phytomining of commercially important metals, hyperaccumulation traits can also be used to improve the nutritional value of food for people living largely on a vegetative diet, which is normally relatively poor in iron and zinc [13]. When considering that about 1.5 billion people worldwide suffer from zinc-deficiency, improving the zinc content in food could clearly aid in solving a public health problem [14].

The most studied metal hyperaccumulator plant species are the Zn, Cd and Ni hyperaccumulator *Thlaspi cearulescens,* Zn, Cd and Ni hyperaccumulator *Arabidopsis halleri* and Cd and Zn hyperaccumulator *Thlaspi praecox* (Fig. **1**) of the Brassicaceae family. Their physiological, morphological and genetic characteristics, and their close relationship to *Arabidopsis thaliana*, the general plant reference species, make them excellent candidates to be the plant heavy metal hyperaccumulation model species [14].

Figure 1: *Thlaspi praecox* Wulfen (photo: K. Vogel-Mikuš).

In this chapter we will briefly review the main findings regarding physiological and molecular mechanisms of metal uptake, transport, accumulation and tolerance in the model hyperaccumulator plant species *Thlaspi caerulescens* and its relative

Cd and Zn hyperaccumulator *Thlaspi praecox*, with special emphasis on hyperaccumulation of cadmium and interactions with symbiotic arbuscular mycorrhizal (AM) fungi. Among the heavy metals, cadmium (Cd) has been considered as one of the most serious metal contaminants adversely affecting both human health and the functioning of ecosystems. Despite being a non-essential element for living organisms, Cd has a very high mobility in soil-plant systems, resulting in accumulation in edible parts of plants (*e.g.,* leaves and grains), therefore it is necessary to take action towards remediation of Cd-contaminated soils [15]. Inoculation with metal-resistant AM fungi can enhance the growth of higher plants through additional supply of water and mineral nutrients as well as by modification of the toxicity of the metal contaminants *via* complexation or precipitation [16, 17]. Metal hyperaccumulator plants of the Brassicaceae family have been for a long time (and still are, by many scientists) considered to be non-mycorrhizal. Recent studies, however, have shown that AM fungi can actually help particular Brassicaceous plant species to better sequester mineral nutrients during the reproductive period in severely heavy metal-polluted soils. This can actually enable survival of the species in such a hostile environment [18-20].

MOLECULAR PHYSIOLOGY OF METAL HYPERACCUMULATION IN *THLASPI*

The genus *Thlaspi* s.l. comprises more than 80 species and it is primarily distributed in Subarctic and north temperate regions in Eurasia and north America. A few species occur in the Andean regions of South America [21]. In the genus *Thlaspi* 23 species are known to hyperaccumulate Ni (*T. alpestre* Jacq., *T. goesingense* Halàcsy, *T. pindicum* Hausskn., *etc.*) [22, 23], 11 species hyperaccumulate Zn (*T. caerulescens, T. praecox* (Bulgaria), *etc.*) [22], one hyperaccumulates Pb (*Thlaspi cepaeifolium* (Wulfen) W.D.J.Koch *ssp. cepaeifolium*) [24] and three hyperaccumulate Cd (*T. caerulescens, T. goesingense and T. praecox*) [25, 26]. Besides on heavy metal enriched/ polluted sites, metal hyperaccumulator *Thlaspi* species can also be found on non-polluted sites. Therefore they are considered to be pseudo-metallophytes.

Metal hyperaccumulation is closely connected to physiological and biochemical adaptations of the root metal uptake, metal transport from the roots to the shoots, metal sequestration and detoxification in shoots or leaves [6, 27-29]. The roots of the Zn/Cd hyperaccumulator *T. caerulescens* preferentially spread towards areas with higher Zn or Cd concentrations in soils, foraging for these two metals [30, 31]. However the mechanisms of sensing metals in soils are more or less unknown.

At the molecular level overexpression of several genes involved in metal uptake into the cells was observed in *T. caerulescens*, namely genes from the ZIP family of metal transporters, such as *ZIP* and *IRT* [11], pointing to the enhanced metal uptake to the roots. Besides enhanced metal uptake into the root symplast, in the metal hyperaccumulator *Thlaspi*, modifications at the level of metal-xylem loading were also observed (Fig. **2**). In non-accumulating plants, for example *T. arvense,* the majority of metals that are taken into the symplast, are then sequestered into the vacuoles. On the other hand in *T. caerulescens*, metals are rather rapidly loaded to the xylem and transferred to the shoots. As a consequence, up to five-times higher Zn concentrations were detected in the xylem sap of *T. caerulescens* compared to *T. arvense* [32, 33]. Studies in *Arabidopsis* showed that HMA2 and HMA4, belonging to the P1B ATP-ase subfamily, play an important role in root-to-shoot translocation of Zn and Cd [34, 35]. HMA2 and HMA4 play a role in the loading or unloading of Zn in the xylem. Expression of both is also observed in the phloem tissue and may indicate a role in the remobilization of Zn from shoots to roots. The sub-cellular localization of HMA2 to the plasma membrane is consistent with transporting Zn into or out of the cells and supports a role for HMA2 in the translocation of Zn within the plant [34]. In the Cd-hyperaccumulator species *T. caerulescens* and *A. halleri*, the expression of *AtHMA4* homologues (*TcHMA4* and *AhHMA4*) showed a higher expression level in both the roots and shoots when compared to related non-accumulators [36-42]. Recently, it was shown that the knockdown of *AhHMA4* in *A. halleri* by RNA interference resulted in decreased accumulation of Zn and Cd in the shoots and increased accumulation of these metals in the roots. These findings indicate that high *HMA4* transcript levels are required for efficient xylem loading of Cd and Zn in hyperaccumulator plant species [43]. Once loaded into the xylem the metals are transported to the shoots either as free hydrated ions, or as stable complexes bound either to free His (as His-Ni and His-Zn complexes), to nicotianamine, organic acids [44-46] or S-ligands like glutathione [45, 47].

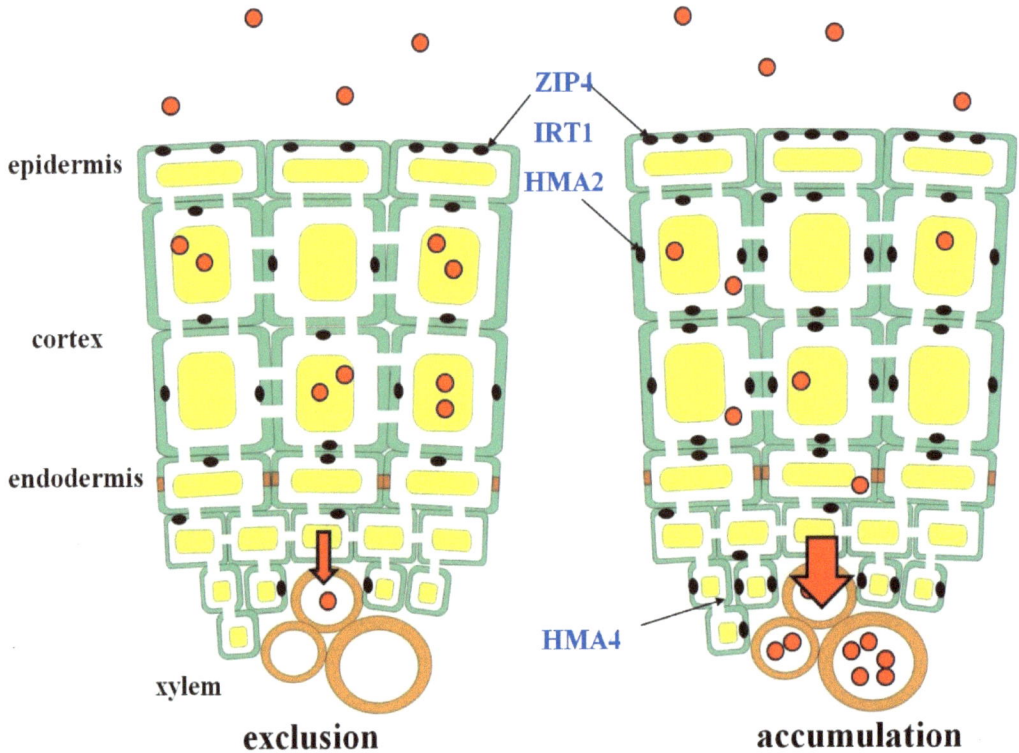

Figure 2: A generalized model for root metal (Zn and Cd) uptake and xylem loading in metal excluder and metal accumulator plants.

As a consequence of enhanced metal uptake and translocation, the majority of Cd and Zn accumulate in the above-ground tissues, primarily the leaves. Since heavy metals are highly toxic and can severely interfere with plant metabolism, especially photosynthesis, an efficient and high-capacity metal detoxification system is required in the leaves of metal hyperaccumulator species in order to enable survival of the plants. When *AhHMA4* from *A. halleri* was expressed in *Arabidopsis thaliana* under the control of the *AhHMA4* promoter, the translocation of Cd to the shoots was increased, but the sensitivity of the transformants to Cd was also increased because of the absence of a shoot detoxification mechanism [43]. Furthermore, when a non-accumulator, *T. perfoliatum*, was grafted onto *T. caerulescens* root stock, translocation of Zn and Cd to the leaves was enhanced, but the leaves showed distinctive symptoms of metal toxicity [48]. These studies clearly demonstrated that functional metal hypertolerance mechanisms are a prerequisite for metal hyperaccumulation.

In metal excluder plants (*e.g., T. arvense*) metals (presented by red circles in Fig. **2**) are mainly sequestered in the vacuoles of the root cells and only to a lesser extent loaded into the xylem. In metal hyperaccumulator plants (*e.g., T. caerulescens*) metal transporters in the roots, namely ZIP4, IRT1, HMA2 and HMA4 are overexpressed, so consequently higher amounts of metals are taken up into the root symplast. In addition, lower amounts of metals are sequestered into the root vacuoles, so the majority of metals are left available in cytosol for further loading into the xylem. HMA4 transporters play an important role in metal loading into the xylem [11, 29, 34].

In leaves of metal hyperaccumulator plants, metals are first unloaded from the xylem into the symplast of vessel-associated parenchyma cells *via* YSL and FRD3 proteins [11]. High metal symplast contents are then detoxified by rapid pumping of metals into the vacuoles, so vacuolar sequestration is considered as the main mechanism for Cd/Zn tolerance (Fig. **3**) [45, 49, 50-54]. The transporter encoded by *TcMTP1* (*ZTP1*) has been proposed to be a major contributor to Zn accumulation in the vacuoles of leaf cells and is therefore involved in the Zn hyperaccumulation demonstrated by this species [55]. *TcMTP1* is expressed mainly in the shoots of *T. caerulescens* and shows a low level of expression in the roots. It also appears that there may be more than one copy of *TcMTP1* in the *Thlaspi* genome [55]. Recently a gene (*TcHMA3*) from a Cd-hyperaccumulator ecotype of *T. caerulescens* (Ganges) that contributes to Cd sequestration and hypertolerance was isolated and characterized [42]. The gene *TcHMA3* encodes a tonoplast-localized transporter specific for Cd, and a high level of expression of this gene is required for Cd hypertolerance in Cd hyperaccumulation. TcHMA3 belongs to the P1B-type ATPase subfamily. Members of this group transport various heavy metal ions from the cytosol out of the cell or to an organelle [56]. In the hyperaccumulator *T. caerulescens*, ecotype Ganges, predominately accumulating Cd, *TcHMA3* is expressed almost to the same exten in the tonoplasts of epidermal and mesophyll leaf cells. In contrast, the AtHMA3 homologue of non-accumulating *Arabidopsis thaliana* is mainly expressed in the hydathodes, guard cells and vascular tissue of *Arabidopsis* leaves [42]. Expression patterns of *TcHMA3* and *AtHMA3* homologues are also in line with Cd distribution patterns observed in leaf tissues of Cd hyperaccumulator and non-accumulating plants. In

Cd hyperaccumulator *T. caerulescens* Cd is mainly accumulated in the vacuoles of epidermal and mesophyll cells [51], while in Cd-non accumulating plants vascular tissues and guard cells are the main storage sites for Cd [42, 57].

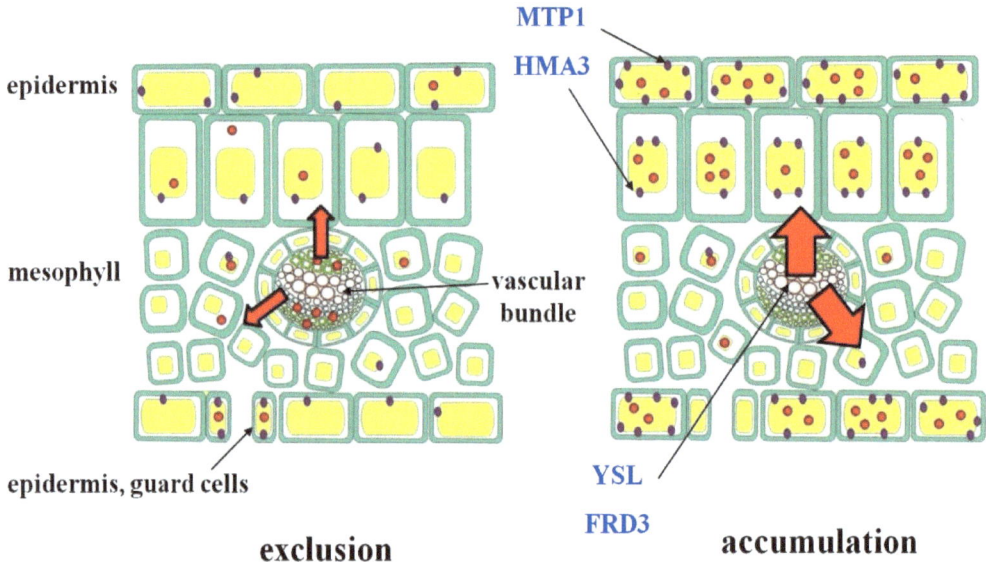

Figure 3: A generalized model of metal trafficking within the leaves of metal excluder and metal hyperaccumulator plants.

In metal excluder plants (*e.g., T. arvense*) metals (presented by red circles in Fig. **3**) are mainly accumulated in vein parenchyma and collenchyma cells and at higher concentrations they can also be found in the mesophyll and guard cells, interfering with their normal functioning. In metal hyperaccumulator plants (*e.g., T. caerulescens*) metals are unloaded from the xylem *via* YSL and FRD3 proteins and then rapidly sequestered in the vacuoles of leaves. Tonoplast transporters MTP1 and TcHMA3 were proposed to play an important role in vacuolar sequestration and metal detoxification in leaves of metal hyperaccumulator plants [11, 42, 57].

In Cd hyperaccumulator *T. praecox,* Cd distribution in leaves of Cd-treated plants was proved to be also dependent on Cd concentrations added to the nutrient solution. At lower Cd concentrations in a nutrient solution, Cd accumulated first in epidermal cells. With increasing Cd concentrations in the nutrient solution and consequently in the leaf tissues, Cd also started to accumulate gradually in the mesophyll cells [58]. In field-collected *T. praecox* plants growing on multi-metal polluted soils (Zn, Cd and Pb) the epidermal cells were predominantly occupied by Zn and Pb, while Cd was mainly stored in the mesophyll [53, 54]. In mixed Cd and Zn *T. praecox* treatments, the epidermal cells apparently favour accumulation of Zn over Cd, but molecular mechanisms that enable distinguishing between metals and selective sequestration are not yet fully understood. It is possible that genes involved in vacuolar sequestration of Cd (*TcHMA3*) and genes involved in vacuolar sequestration of Zn (*MTP1*) may be differentially expressed in the mesophyll and epidermal tissues. This might result in differential affinity of epidermal and mesophyll cells for both metals present in plant tissues at the same time. However further studies are needed to confirm this.

A crucial role for efficient vacoular sequestration of Cd and Zn in hypertolerance was confirmed also through studies of Cd and Zn ligands in plant tissues. In *T. caerulescens* metal sequestration mainly depends on the age of the leaves [45]. In older leaves Zn and Cd were mainly bound to oxygen ligands (organic acids such as malate and citrate), while in younger leaves Cd was manly bond to S ligands (possibly phytochelatins, metallotioneins and Cys-rich proteins), probably because in young leaves the vacuoles were not yet fully developed. In addition, part of the Cd and Zn was also bound to free His [45]. Coexistence of oxygen and sulphur Cd ligands in leaves was also confirmed in field-collected *T. praecox* using Cd-K edge extended X-ray absorption fine structure [47]. Both mentioned studies were performed on bulk plant materials, so there is still a lack of information on spatial distribution of Cd-ligand complexes at the cellular and tissue level. A study which involved isolation of epidermal, mesophyll and vein tissues in field-collected Cd hyperaccumulator *T. praecox* showed that there was no significant difference in the involvement of Cd ligands between epidermal and mesophyll tissues, since in both cases oxygen ligands strongly prevailed [47]. In veins, however, Cd was mainly bound to sulphur ligands (57% of S and 43% of O ligands), probably on the account of Cd stored in phloem tissues which could not be separated from xylem during sample preparation [47]. Phloem is usually very rich with different organic compounds such as amino-acids, free -SH group of cysteine, glutathione and other thiol compounds, which could serve as Cd ligands [47].

In *T. praecox* a significant amount of Cd was also found to be stored in the cell walls [53], which can also act as a very important, but often overlooked Cd detoxification mechanism. According to "egg box" hypothesis [59], cell walls that are rich in pectins can provide numerous hydroxylic and carboxylic groups with high affinity for binding metals. Entrapping of metal ions into the cell wall components of fully differentiated cells can represent an important way of detoxifying the symplast and preventing interference of metals with vitally important metabolic pathways [47, 53].

Compared to the plant species that are unable to tolerate excessive soil metal concentrations, plants that inhabit metal-polluted sites are also adapted at the reproductive stage. Metals are transported to the developing seeds by the phloem, so metal uptake depends basically on the phloem loading within the leaves [60]. In metal-tolerant plants efficient metal sequestration mechanisms maintain low free metal concentrations in symplast, thus limiting metal phloem loading and transport from the leaves to the seeds [33, 61], for example, Zn in seeds of *T. praecox* [62]. The seeds of metal-tolerant plants thus contain lower metal concentrations than any other plant parts, enabling the embryos to survive in metal-polluted environments [23, 27]. The seed coat and/or endosperm represent another barrier preventing metal accumulation in embryonic tissues [27, 63, 64]. Only little is known about the accumulation and distribution of metals in seeds of plants from the genus *Thlaspi*. In the Ni hyperaccumulator *T. pindicum* Ni is mainly accumulated in the seed-coat micropylar region, while in the embryo Ni is mainly found in the epidermis of cotyledons [23]. On the contrary in *T. caerulescens*, so as *T. praecox*, Zn and Cd were found to accumulate in the embryonic tissues [62, 65]. In *T. praecox,* concentrations of Cd were found to be at an order of magnitude higher than that of Zn. Further research in field-collected plants indicated that Zn is probably more efficiently sequestered in leaves and much less mobile than Cd [53, 54]. In *T. praecox* seeds almost two thirds of the Cd ligands were thiol groups (Cd-S-C-). In addition, a coordination to phosphate groups *via* bridging oxygens (Cd-O-P-), as for phytate was observed, indicating that Cd is behaving similarly as Zn, which also bound to phytate when stored in seeds [47]. The faith of Cd during germination still needs to be elucidated in order to better understand the mechanisms which enable survival of the embryo during germination in the presence of this highly toxic Cd load.

INTERACTIONS OF METAL HYPERACCUMULATOR *THLASPI* WITH ARBUSCULAR MYCORRHIZAL FUNGI

Arbuscular mycorrhiza (AM) is a ubiquitous symbiosis between fungi in the order of *Glomeromycota* and plant roots [66]. More than 90% of plants are known to form this type of symbiosis, where the plants provide carbohydrates to the fungi in exchange for mineral nutrients, especially phosphorus [67].

In plants growing on highly metal-polluted sites, symbiosis with suitable arbuscular mycorrhizal (AM) fungi may be seen as an additional tolerance mechanism [28, 68]. AM fungi were shown to protect plants from metal toxicity, which is connected mainly to fungal metal tolerance mechanisms, such as metal sequestration in the cell walls and vacuoles of the intra- and inter-radical mycelium [69-73].

Metal hyperaccumulator plants of the genus *Thlaspi* were (and still are, by many scientists) regarded as non-mycorrhizal [3, 74-76], since the genus belongs to the Brassicaceae family, with a well-established non-mycorrhizal status [77]. The lack of mycorrhizal symbiosis was mainly attributed to the differences in phosphate acquisition/ scavenging systems compared with mycorrhizal species [78], the formation of fungitoxic/ fungistatic breakdown products from glucosinolates [79], or the lack of communication signals between the symbionts [79]. The findings of AM fungal colonization in Brassicaceous plant species like *Biscutella laevigata* [80] and *Thlaspi* sp. [18, 26, 81, 82] have opened new questions about the development and the role of AM symbiosis in Brassicaceae, especially in those inhabiting metal-polluted sites or even hyperaccumulating metals.

T. praecox from metal-polluted site in Slovenia was found to be distinctly colonised by AM fungi [26, 82], but a low incidence of AM colonization, with an average mycorrhizal frequency (F as %) of approximately 20%, was observed when compared to the other plant species [26, 82]. At non-polluted sites F of up to 70% was observed [26]. Screening of AM fungal colonization in different species of the genus *Thlaspi* revealed that the species growing in an active meadow community (*e.g., T. praecox, T. sylvestre = T. caerulescens* (Banjščice) and *T.*

montanum) were distinctly colonised by AM fungi, while only fungal hyphae were observed in the roots of the species collected at metal-polluted sites (*e.g., T. caerulescens, T. goesingense, T. calaminare* and *T. cepaeifolium*) [81]. There could be several reasons for the difference in the AM fungal colonization levels among *Thlaspi* species and/ or populations growing at metal-polluted and non-polluted sites. A part of the energy, which the plants could invest in AM symbiosis in metal-polluted environments, is spent on metal tolerance mechanisms [6]. Therefore the plants may restrict the level of AM colonisation to survive at metal-polluted sites [82].

The selected species from the genus *Thlaspi* were found to be colonised by the common *Glomus intraradices* AM fungi [81]. However, none of the DNA sequences obtained matched any other deposited in databanks, indicating the existence of a species continuum in the *G. intraradices* clade [81]. Several attempts were made but failed to establish AM symbiosis in *T. praecox* under controlled conditions using isolates of *G. intraradices* [81]. Inoculum composed of an indigenous mixture [83], the combination closest to that of the natural environment, was therefore used to increase the possibility of meeting compatible symbionts [18]. The development of AM symbiosis was observed only during the flowering stage of *T. praecox*, after a longer exposure of the plants to low temperatures (vernalization) while at the vegetative stage the plants were not colonized [18]. AM fungi that succeeded to colonize greenhouse grown plants belonged to *Glomus* species (Glomeromycota) [20]. In addition plants were colonized also by putative dark septate fungi *Phialophora verrucosa* and *Rhizoctonia* sp. and by some other fungi from Asco- and Basidiomycota, that are known to associate with plants, namely *Capnobotryella* sp., *Penicillium brevicompactum, Rodotorula aurantiaca* and *Rodotorula slooffiae* [20].

Despite the fact that AM symbiosis was observed only during flowering stage, an increase in mineral nutrient uptake, *e.g.,* phosphorus and sulphur, was observed in inoculated plants, indicating functional exchange of nutrients and probably also carbohydrates between the symbionts. *Paris* type of AM fungal morphology with limited development of arbuscules was observed under controlled conditions [18], indicating hyphae and coils may also play an important role in the nutrient exchange [67]. The level of mycorrhizal frequency (F%) and mycorrhizal intensity (M%) of *T. praecox* increased with Cd and Pb concentrations in the soil under greenhouse conditions [18]. Colonised plants showed decreased Cd uptake to the roots and shoots and decreased Zn uptake to the roots. In addition, the Pb uptake strategy of *T. praecox* was changed in the presence of AM fungi [18]. The changes in metal uptake strategies of *T. praecox* plants colonized by AM fungi therefore indicate the protective role of AM fungi in metal-polluted soils [18, 69-73]

Improved nutrient uptake during the reproductive period can result in the production of high quality seed mineral reserves [84-86], which is important for early plant establishment in nutrient-limited environments typical of limestone and metal-polluted sites [3, 60, 87-89]. Among mineral nutrients in AM-inoculated plants the most significant was the improvement of P and S uptake [18]. The increased need for metal tolerance mechanisms during the reproductive period is further supported by the finding that *T. praecox* seeds hyperaccumulate significant amounts of Cd, which can negatively affect seed biomass, germination potential and viability [62]. Since Cd in the seeds is bound to sulphur and phosphorus ligands, increased S and P uptake *via* AM fungi could improve Cd detoxification in embryos and therefore enable their survival.

The non-mycorrhizal status of Brassicaceae can be mainly attributed to defence substances, glucosinolates (GS) and their hydrolysis products, the presence or absence of lectins, and/or the structural-chemical properties of the root cell walls that hinder or inhibit fungal growth [90]. Especially the presence and/or absence of particular GS in plant roots have been suggested to largely contribute to the host/ non-host AM character of GS-containing plant species [79]. Because most of the non-AM colonizing plants contained gluconasturtiin (2-phenylethyl GS) in roots, this GS, but not others, has been proposed to act as a general AM inhibitory factor [79]. In our study [20] where the level of AM colonization and profiles and contents of glucosinolates were monitored during the life cycle of field grown *T. praecox*, no gluconasturtiin was found in plants roots, so as in other GS-containing plant species that are susceptible to AM colonization. In addition, a distinct change of GS profile and a decrease of total GS concentration in roots coincided with the peak of AM colonization level found in *T. praecox* during flowering and seeding period. The presence of glucotropaeolin and absence of glucobrassicanapin was also observed during the reproductive phases of *T. praecox,* indicating a possible role of these two glucosinolates in communication between the symbionts and formation of AM symbiosis. Quantities of glucotropaeolin, the main GS of the AM-host *Tropaeolum majus*, have already been shown to increase during AM colonization [79, 91], while glucobrassicanapin, on the other hand, has been found only in AM non-host *Brassica napus* [79]. Thus, the pattern of GS in *T. praecox* roots, *i.e.*, the presence and absence of specific GS, is likely to affect AM formation, as some GS may inhibit and others may help to induce AM formation. In addition to GS profiles and/or contents, plant mineral demand, seasonal dynamics of mycorrhization [19], and/or diverse signalling molecules [92] may all contribute to AM colonization in *T. praecox* [20].

CONCLUSIONS

In the past two decades studies of the mechanism of metal uptake, transport, tolerance and accumulation in model hyperaccumulator plant species using molecular approaches, as well as modern techniques for metal localization, such as micro-PIXE, and determination of the metal local environment using extended X-ray absorption fine structure and X-ray absorption near edge structure, have partly revealed the basic features of the physiology of metal hyperaccumulation and tolerance in order to enhance the use of metal hyperaccumulator plants in phytoremediation. However there is still a lot of issues left to be studied, especially in the field of molecular mechanism of metal hyperaccumulator plant - microbe interactions, because soil microorganisms can significantly alter metal and mineral nutrient uptake.

ACKNOWLEDGEMENTS

Research on *Thlaspi preacox* was supported by Plant biology program group P1-0212. Collaborators Paula Pongrac, Peter Kump, Marijan Nečemer, Iztok Arčon, Primož Pelicon, Primož Vavpetič, Matevž Likar, Sliva Sonjak, Marjana Regvar, Charlotte Poschenrieder, Juan Barcelo, Jolanta Mesjasz-Przybylowicz and Wojtek Przybylowicz are acknowledged.

CONFLICT OF INTEREST

The author(s) confirm that this chapter content has no conflict of interest.

REFERENCES

[1]   Barceló J, Poschenrieder C. Phytoremediation: principles and perspectives. Contribut Sci 2003; 2: 333-444.

[2]   Ensley BD. Rationale for Use of Phytoremediation. In: Raskin I, Ensley BD, Eds. Phytoremediation of Toxic Metals, Using Plants to Clean up the Environment, 1st Ed. New York, John Wiley, Sons, Inc, 2000; pp. 53-70.

[3]   Ernst WHO. Phytoextraction of mine wastes - options and impossibilities. Chemie der Erde 2005; 65 S1: 29-42.

[4]   Timofeev-Resovsky EA, Agafonov BM, Timofeev-Resovsky NV. Fate of radioisotopes inaquatic environments (In Russian). Proceed Biolog Instit USSR Acad Sci 1962; 22: 49-67.

[5]   Prasad MNV, Freitas HMO. Metal hyperaccumulation in plants - Biodiversity prospecting for phytoremediation technology. E J Biotechnol 2002; 3: 285-321.

[6]   Baker AJM. Metal tolerance. New Phytol 1987; 106 S: 93-111.

[7]   Baker AJM. Accumulators and excluders - strategies in the response of plants to heavy metals. J Plant Nutrit 1981; 3: 643-54.

[8]   Shen ZG, Zhao FJ, McGrath SP. Uptake and transport of zinc in the hyperaccumulator *Thlaspi caerulescens* and the nonhyperaccumulator *Thlaspi ochroleucum*. Plant Cell Environ 1997; 20: 898-906.

[9]   Brooks RR, Lee J, Reeves RD, Jaffre T. Detection of nickeliferous rocks by analysis of herbarium specimens of indicator plants. J Geochem Explorat 1977; 7: 49-57.

[10]   Reeves RD, Baker AJM. Metal Accumulating Plants. In: Raskin I, Ensley BD, Eds. Phytoremediation of Toxic Metals, Using Plants to Clean up the Environment, 1st Ed. New York, John Wiley & Sons Inc, 2000: pp. 193-229.

[11]   Verbruggen N, Hermans C, Schat H. Molecular mechanisms of metal hyperaccumulation in plants. New Phytol 2009; 181: 759-76.

[12]   Baker AJM., Whiting, SN. In search of the holy grail - a further step in understanding metal hyperaccumulation? New Phytol 2002; 155: 1-4.

[13]   Grusak MA. Enhancing mineral content in plant food products. J Am Col Nutrit 2002; 21: 178S-83S.

[14]   Assunção AG, Schat H, Aarts MG. *Thlaspi caerulescens*, an attractive model species to study heavy metal hyperaccumulation in plants. New Phytol 2003; 159: 351-60.

[15]   Ji P, Tieheng T, Song Y, Ackland ML, Liu Y. Strategies for enhancing the phytoremediation of cadmium-contaminated agricultural soils by *Solanum nigrum* L, Environ Pollut 2010;159: 762-8.

[16]   Gonzalez-Chavez C, Harris PJ, Dodd J, Meharg AA. Arbuscular mycorrhizal fungi confer enhanced arsenate resistance to *Holcus lanatus*. New Phytol 2002; 155: 163-71.

[17]   Kaldorf M, Kuhn AJ, Schröder WH, Hildebrand U, Bothe H. Selective element deposits in maize colonized by a heavy metal tolerance conferring arbuscular mycorrhizal fungus. J Plant Physiol 1999; 154: 718-28.

[18] Vogel-Mikuš K, Pongrac P, Kump P, Nečemer M, Regvar M. Colonization of a Zn, Cd and Pb hyperaccumulator *Thlaspi praecox* Wulfen with indigenous arbuscular mycorrhizal fungal mixture induces changes in heavy metal and nutrient uptake. Environ Pollut 2006; 139: 362-71.

[19] Pongrac P, Vogel-Mikuš K, Kump P, Nečemer M, Tolrá R, Poschenrieder C, Barceló J, Regvar M. Changes in elemental uptake and arbuscular mycorrhizal colonisation during the life cycle of *Thlaspi praecox* Wulfen. Chemosphere 2007; 69: 1602-9.

[20] Pongrac P, Sonjak S, Vogel-Mikuš K, Kump P, Nečemer M, Regvar M. Roots of Metal Hyperaccumulating Population of *Thlaspi praecox* (Brassicaceae) Harbour Arbuscular Mycorrhizal and Other Fungi Under Experimental Conditions. Int J Phytoremediat 2009: 11: 347-59.

[21] Koch M, Mummenhoff K, Hurlka H. Systematics and evolutionary history of heavy metal tolerant *Thlaspi caerulescens* in Western Europe. Biochem System Ecol 1998; 26: 823-38.

[22] Brooks RR. General Introduction. In: Brooks RR, Ed. Plants That Hyperaccumulate Heavy Metals, 2nd Ed. Wallingford, CAB international, 1998; pp. 1-14.

[23] Psaras GK, Manetas Y. Nickel localization in Seeds of the metal hyperaccumulator *Thlaspi pindicum* Hausskn. Ann Bot 2001; 88: 513-6.

[24] Reeves RD, Brooks RR. Hyperaccumulation of lead and zinc by two metallophytes from mining areas of central Europe. Environ Pollut Series A, Ecol Biol 1983; 31: 277-85.

[25] Lombi E, Zhao FJ, Dunham SJ, McGrath SP. Cadmium accumulation in populations of *Thlaspi caerulescens* and *Thlaspi goesingense*. New Phytol 2000; 145: 11-20.

[26] Vogel-Mikuš K, Drobne D, Regvar M. Zn, Cd and Pb accumulation and arbuscular mycorrhizal colonization of pennycress *Thlaspi praecox* Wulf. Brassicaceae from the vicinity of a lead mine and smelter in Slovenia. Environ Pollut 2005; 133: 233-42.

[27] Ernst WHO, Verkleij JAC, Schat H. Metal tolerance in plants. Acta Bot Neerl 1992; 41: 229-48.

[28] Hall JL. Cellular mechanisms for heavy metal detoxification and tolerance. J Exp Bot 2002; 366: 1-11.

[29] Milner M, Kochian LV. Investigating heavy-metal hyperaccumulation using *Thlaspi caerulescens* as a model system. Ann Bot 2008; 102:3-13.

[30] Whiting SN, Leake JR, McGrath SP, Baker AJM. Positive responses to Zn and Cd by roots of the Zn and Cd hyperaccumulator *Thlaspi caerulescens*. New Phytol 2000; 145: 199-210.

[31] Haines BJ. Zincophilic root foraging in *Thlaspi caerulescens*. New Phytol 2002; 155: 363 - 72.

[32] Lasat MM, Baker AJM, Kochian LV. Physiological characterisation of root Zn^{2+} absorption and translocation to shoots in Zn hyperaccumulator and nonaccumulator species of *Thlaspi*. Plant Physiol 1996; 112: 1715-22.

[33] Lasat MM, Baker AJM, Kochian LV. Altered Zn compartmentation in root symplasm and stimulated Zn absorption into the leaf as mechanisms involved in Zn hyperaccumulation in *Thlaspi caerulescens*. Plant Physiol 1998; 118: 875-83.

[34] Hussain D, Haydon MJ, Wang Y, *et al.* P-type ATPase heavy metal transporters with roles in essential zinc homeostasis in *Arabidopsis*. Plant Cell 2004; 16: 1327-39.

[35] Wong CKE, Cobbett CS. HMA P-type ATPases are the major mechanism for root-to-shoot Cd translocation in *Arabidopsis thaliana*. New Phytol 2009; 181: 71-8.

[36] Becher M, Talke IN, Krall L, Krämer U. Cross-species microarray transcript profiling reveals high constitutive expression of metalhomeostasis genes in shoots of the zinc hyperaccumulator *Arabidopsis halleri*. Plant J 2004; 37, 251-268.

[37] Bernard C, Roosens N, Czernic P, Lebrun M, Verbruggen, N. A novel CPx-ATPase from the cadmium hyperaccumulator *Thlaspi caerulescens*. FEBS Lett 2004; 569: 140-8.

[38] Papoyan A, Kochian LV. Identification of *Thlaspi caerulescens* genes that may be involved in heavy metal hyperaccumulation and tolerance. Characterization of a novel heavy metal transporting ATPase. Plant Physiol 2004; 136: 3814-23.

[39] Hammond JP, Bowen HC, White PJ, *et al.* A comparison of the *Thlaspi caerulescens* and *Thlaspi arvense* shoot transcriptomes. New Phytol 2006; 170: 239-60.

[40] Talke IN, Hanikenne M, Krämer U. Zinc-dependent global transcriptional control, transcriptional deregulation, and higher gene copy number for genes in metal homeostasis of the hyperaccumulator *Arabidopsis halleri*. Plant Physiol 2006; 142: 148-67.

[41] Courbot M, Willems G, Motte P, *et al.* A major quantitative trait locus for cadmium tolerance in *Arabidopsis halleri* colocalizes with HMA4, a gene encoding a heavy metal ATPase. Plant Physiol 2007; 144: 1052-65.

[42] Ueno D, Milner MJ, Yamaji N, *et al.* Elevated expression of TcHMA3 plays a key role in the extreme Cd tolerance in a Cd-hyperaccumulating ecotype of *Thlaspi caerulescens*. Plant J 2011; 66: 852-86.

[43] Hanikenne M, Talke IN, Haydon MJ, *et al.* Evolution of metal hyperaccumulation required *cis*-regulatory changes and triplication of HMA4. Nature 2008; 453: 391-5.

[44] Krämer U, Cotter-Howells JD, Charnock JM, Baker AJM, Smith JAC. Free histidine as a metal chelator in plants that accumulate nickel. Nature 1996; 378: 635-8.

[45] Küpper H, Mijovilovich A, Klaucke-Mayer W, Kroneck PHM. Tissue and age-dependent differences in the complexation of cadmium and zinc in the cadmium/zinc hyperaccumulator *Thlaspi caerulescens* Ganges ecotype revealed by X-ray absorption spectroscopy. Plant Physiol 2004: 134: 748-57.

[46] Pilon-Smits E. Phytoremediation. Ann Rev Plant Biol 2005; 56: 15-39.

[47] Vogel-Mikuš K, Arčon I, Kodre A. Complexation of cadmium in seeds and vegetative tissues of the cadmium hyperaccumulator Thlaspi praecox as studied by X-ray absorption spectroscopy. Plant Soil 2010; 331:439-51.

[48] de Guimarães MA, Gustin JL, Salt DE. Reciprocal grafting separates the roles of the root and shoot in zinc hyperaccumulation in *Thlaspi caerulescens*. New Phytol 2009; 184: 323-9.

[49] Küpper H, Zhao FJ, McGrath SP. Cellular compartmentation of zinc in leaves of the hyperaccumulator *Thlaspi caerulescens*. Plant Physiol 1999; 119: 305-11.

[50] Cosio C, DeSantis L, Frey B, Diallo S, Keller C. Distribution of cadmium in leaves of *Thlaspi caerulescens*. J Exp Bot 2005; 56: 765-75.

[51] Ma JF, Ueno D, Zhao FJ, McGrath SP. Subcellular localisation of Cd and Zn in the leaves of a Cd hyperaccumulating ecotype of *Thlaspi caerulescens*. Planta 2005; 220:731-36.

[52] Ueno D, Ma JF, Iwashita T, Zhao FJ, McGrath SP. Identification of the form of Cd in the leaves of a superior Cd-accumulating ecotype of *Thlaspi caerulescens* using 113Cd-NMR. Planta 2005; 221:928-36.

[53]   REGVAR, Marjana, EICHERT, Diane, KAULICH, Burkhard, GIANONCELLI, Alessandra, PONGRAC, Paula, VOGEL-MIKUŠ, Katarina. Biochemical characterization of cell types within leaves 5 of metal-hyperaccumulating Noccaea praecox (Brassicaceae). Plant soil. [Print ed.], 2013 [in press], doi: 10.1007/s11104-013-1768-z.

[54]   Vogel-Mikuš K, Simčič J, Pelicon P, *et al.* Comparison of essential and non-essential element distribution in leaves of the Cd/Zn hyperaccumulator *Thlaspi praecox* as revealed by micro-PIXE. Plant Cell Environ 2008a; 31:1484-96.

[55]   Vogel-Mikuš K, Regvar M, Mesjasz-Przybyłowicz J, *et al.* Spatial distribution of cadmium in leaves of metal hyperaccumulating *Thlaspi praecox* using micro-PIXE. New Phytol 2008; 179:712-21.

[56]   Assunção AGL, Da CostaMartins P, De Folter S, Vooijs R, Schat H, Aarts, MGM. Elevated expression of metal transporter genes in three accessions of the metal hyperaccumulator *Thlaspi caerulescens*. Plant Cell Environ 2001; 24: 217-26.

[57]   Williams LE, Mills RF. P1B-ATPases - an ancient family of transition metal pumps with diverse functions in plants. Trends Plant Sci 2005; 10: 491-502.

[58]   Vollenweider P, Cosio C, Günthardt-Goerg MS, Keller C. Localization and effects of cadmium in leaves of a cadmium-tolerant willow (*Salix viminalis* L.). Part II Microlocalization and cellular effects of cadmium. Environ Exp Bot 2006; 58: 25-40.

[59]   Pongrac P, Vogel-Mikuš K, Vavpetič P, *et al.* Cd induced redistribution of elements within leaves of the Cd/Zn hyperaccumulator *Thlaspi praecox* as revealed by micro-PIXE. Nuclear Instruments and Methods Physics Res B Beam Interaction Materials Atoms 2010; 268: 2205-10.

[60]   Grant GT, Morris ER, Rees DA, Smith PJC, Thom D. Biological interactions between polysaccharides and divalent cations: the egg-box model. FEBS Lett 1973; 32: 195-8.

[61]   Marschner H. Mineral Nutrition of Higher Plants, 2nd Ed. Academic press, London, United Kingdom, 1995.

[62]   Wojcik M, Vangronsveld J, Haen JD, Tukiendorf A. Cadmium tolerance in *Thlaspi caerulescens* II. Localization of cadmium in *Thlaspi caerulescens*. Environ Exp Bot 2005; 53: 163-71.

[63]   Vogel-Mikuš K, Pongrac P, Kump P, *et al.* Localisation and quantification of elements within seeds of Cd/Zn hyperaccumulator *Thlaspi praecox* by micro-PIXE. Environ Pollut 2007; 147:50-9.

[64]   Mesjazs-Przybylowicz J, Grodzinska K, Przybylowicz WJ, Godzik B, Szarek-Lukaszewska G. Nuclear microprobe studies of elemental distribution in seeds of *Biscutella laevigata* L. from zinc wastes in Olkusz, Poland. Nuclear Instruments and Methods in Physics Res Section B- Beam Interactions with Materials and Atoms 2001; 181: 634-9.

[65]   Bhatia NP, Orlic I, Siegele R, Ashwath N, Baker AJM, Walsh KB. Elemental mapping using PIXE shows the main pathway of nickel movement is principally symplastic within the fruit of the hyperaccumulator *Stackhousia tryonii*. New Phytol 2003; 160: 479-88.

[66]   Kachenko AG, Bhatia NP, Siegele R, Walsh KB, Singh B. Nickel, Zn and Cd localisation in seeds of metal hyperaccumulators using mu-PIXE spectroscopy. Nuclear Instruments & Methods in Physics Res Sect B-Beam Interactions with Materials and Atoms 2009; 267: 2176-80.

[67] Schüßler A, Gehrig H, Schwarzott D, Walker C. Analysis of partial Glomales SSU rRNA genes: implications for primer design and phylogeny. Mycol Res 2001; 105: 5-15.

[68] Smith SE, Read DJ. Mycorrhizal Symbiosis, 2nd Ed. Academic Press, London, United Kingdom 1997.

[69] Turnau K, Miszalski Z, Trouvelot A, Bonfante P, Gianinazzi S. *Oxalis acetosella* as a Monitoring Plant on Highly Polluted Soils. In: Azcon-Aguilar C, Barea JM, Eds. Mycorrhizas in Integrated Systems, from Genes to Plant Development. Proceed Fourth Europ Symp Mycorrhizas. COST Ed. Brussels, Luxemburg, European Commission; 1996; pp. 483-6.

[70] Gildon A, Tinker PB. Interactions of vesicular-arbuscular mycorrhizal infection and heavy metals in plants. I. The effects of heavy metals on the development of vesicular-arbuscular mycorrhizas. New Phytol 1983; 95: 247-61.

[71] Dehn B, Schüepp H. Influence of VA mycorrhizae on the uptake and distribution of heavy metals in plants. Agri, Ecosyst Environ 1989; 29: 79 -83.

[72] Hetrick BAD, Wilson GWT, Figge DAH. The influence of mycorrhizal symbiosis and fertilizer amendments on establishment of vegetation in heavy metal mine spoils. Environ Pollut 1994; 86: 171-9.

[73] Hildebrandt U, Kaldorf M, Bothe H. The zinc violet and its colonization by arbuscular mycorrhizal fungi. J Plant Physiol 1999; 154: 709-17.

[74] Chen BD, Tao HQ, Christie P, Wong MH. The role of arbuscular mycorrhiza in zinc uptake by red clover growing in a calcareous soil spiked with various quantities of zinc. Chemosphere 2003; 50: 839-46.

[75] Leyval C, Turnau K, Haselwandter K. Effect of heavy metal pollution on mycorrhizal colonization and function, physiological, ecological and applied aspects. Mycorrhiza 1997; 7: 139-53.

[76] Pawlowska TE, Blaszkowski J, Rühling Å. The mycorrhizal status of plants colonizing a calamine spoil mound in southern Poland. Mycorrhiza 1996; 6: 499-505.

[77] Coles KE, David JC, Fisher PJ, Lappin-Scott HM, Macnair MR. Solubilisation of zinc compounds by fungi associated with the hyperaccumulator *Thlaspi caerulescens*. Bot J Scot 2001; 51: 237-47.

[78] Harley JL, Harley EL. A check-list of mycorrhiza in the British flora. New Phytol 1987; 105S: 1-102.

[79] Murley VR, Theodorou ME, Plaxton WC. Phosphate starvation - inducible pyrophosphate - dependent phosphofructokinase occurs in plants whose roots do not form symbiotic association with mycorrhizal fungi. Physiol Plant 1998; 103: 405-14.

[80] Vierheilig H, Bennett R, Kiddle G, Kaldorf M, Ludwig-Müller J. Differences in glucosinolate patterns and arbuscular mycorrhizal status of glucosinolate- containing plant species. New Phytol 2000; 146: 343-52.

[81] Orlowska E, Zubek Sz, Jurkiewitcz A, Szarek-Lukaszeewska G, Turnau K. Influence of restoration on arbuscular mycorrhiza of *Biscutella leavigata* L. Brassicaceae and *Plantago lanceoolata* Plantaginaceae from calamine spoil mounds. Mycorrhiza 2002; 12: 1-17.

[82] Regvar M, Vogel K, Irgel N, Wraber T, Hildebrandt U, Wilde P, Bothe H. Colonization of pennycresses *Thlaspi sp.* of the Brassicaceae by arbuscular mycorrhizal fungi. J Plant Physiol 2003; 160: 615-26.

[83] Regvar M, Vogel-Mikuš K, Kugonič N, Turk B, Batič F. Vegetational and mycorrhizal successions at a metal polluted site - indications for the direction of phytostabilisation? Environ Pollut 2006: 144: 976-84.

[84] Regvar M, Groznik N, Goljevšček N, Gogala N. Diversity of arbuscular mycorrhizal fungi at various differentially managed ecosystems in Slovenia. Acta Biol Slovenica 2001; 44: 27-34.

[85] Koide R, Lu X. Mycorrhizal Infection of Wild Oats, Parental Effects on Offspring Nutrient Dynamics, Growth and Reproduction. In: Read DJ, Lewis DH, Fitter AH, Alexander IJ, Eds. Mycorrhiza in Ecosystems, 1st Ed. Walingford, CAB International, 1992; pp. 55-8.

[86] Allsopp N, Stock WD. Mycorrhizas, Seed Size and Seedling Establishment in a Low Nutrient Environment. In: Read, DJ, Lewis, DH, Fitter, AH, Alexander, IJ, Eds. Mycorrhiza in Ecosystems. Wallingford, CAB International, 1992; pp. 59-64.

[87] Jacobsen I, Smith SE, Smith FA. Function and diversity of arbuscular mycorrhizae in carbon and mineral nutrition. In: van der Heijden MGA, Sanders IR, Eds. Mycorrhizal Ecology, 1st Ed. Berlin, Heidelberg, New York, Springer-Verlag, 2002; pp. 75-92.

[88] Marschner H. Role of root growth, arbuscular mycorrhizal, and root exudates for the efficiency in nutrient acquisition. Field Crops Res 1998: 56: 203-7.

[89] Tyler G, Zohlen A. Plant seeds as mineral nutrient resource for seedlings: a comparison of plants from calcareous and silicate soils. Ann Bot 1998; 81: 455-9.

[90] Adriano DC. Trace Elements in Terrestrial Environments, Biochemistry, Bioavailability and Risk of Metals, 2nd Ed. Springer-Verlag, New York, Berlin, Heidelberg 2001.

[91] Glenn MG, Chew FS, Williams PH. Influence of glucosinolate content of *Brassica* (Cruciferae) roots on growth of vesicular-arbuscular mycorrhizal fungi. New Phytol 1988; 110: 217-25.

[92] Ludwig-Müller J, Bennett RN, Garcia-Garrido JM, Piché Y, Vierhelig H. Reduced arbuscular mycorrhizal root colonization in *Tropaeolum majus* and *Carica papaya* after jasmonic acid application cannot be attributed to increased glucosinolate levels. J Plant Physiol 2002; 159:517-23.

Send Orders for Reprints to reprints@benthamscience.net

CHAPTER 6

From *Arabidopsis thaliana* to Genetic Engineering for Enhanced Phytoextraction of Soil Heavy Metals

David W.M. Leung[*]

School of Biological Sciences, University of Canterbury, Private Bag 4800, Christchurch 8100, New Zealand

Abstract: Soils contaminated with toxic levels of heavy metals present serious public health hazards. A potentially green, environmentally friendly and sustainable technology is phytoextraction of soil heavy metals. Naturally occurring metal hyperaccumulating plants have been found in specific metal enriched habitats but they are not suitable for practical phytoextraction purposes. Genetic engineering is a powerful technology to improve the phytoextraction potential of non-hyperaccumulating plants that have many other desirable attributes over the naturally occurring metal hyperaccumulating plants. In this chapter, the utility of *Arabidopsis thaliana*, a non-hyperaccumulator of heavy metals, to gain novel insights into how the different metal resistance-related genes might operate when plants are exposed to excess soil toxic metals was discussed. It is concluded that use of *A. thaliana* in this way is very useful to genetic engineering for enhanced phytoextraction of soil heavy metals by non-hyperaccumulating plants.

Keywords: Acyl-CoA-binding proteins (ACBPs), *Arabidopsis desaturase 2* gene, *Arabidopsis Ethylene-Insensitive* 2 gene, AtABCC1, AtABCC2, *AtATM3*, *AtPDR8*, *AtPRD12*, *Brassica juncea*, efflux pumping, ethylmethane sulfonate (EMS) mutagenesis, genetic engineering, *GSH1*, health hazard, knockout plants, metal resistance, mutant isolation, mutant screen, phytochelatin synthase (PC synthase), plasma membrane, transgenic plants.

INTRODUCTION

Arabidopsis thaliana is a valuable experimental model system in many different areas of plant biology [1]. For example, the availability of several thousand mutants and various well characterized cloned genes have been used to generate

*Address correspondence to David W.M. Leung:** School of Biological Sciences, University of Canterbury, Private Bag 4800, Christchurch 8100, New Zealand; Tel: 64 3 3642650; Fax: 64 3 3642590; E-mail: david.leung@canterbury.ac.nz

an ever-expanding understanding of many different processes associated with plant growth and development [2]. Besides, many studies on *A. thaliana* have also been used to show how plants react in response to exposure to biologically non-essential heavy metals such as Cd and Pb [3].

Metal hyperaccumulating plants are attractive for phytoremediation purposes, particularly phytoextraction [4]. Only a small number of naturally occurring metal hyperaccumulating plants, which could accumulate several hundred fold more metals than most other plants without suffering toxicity, have been identified from metal-polluted sites. Thus, it is of interest to develop strategies to increase phytoextraction potential of selected non-hyperaccumulating plants through genetic engineering techniques [5]. To this end, there are some extensive efforts to characterize the metal hyperaccumulation traits at the molecular level in a few of the naturally occurring metal hyperaccumulators [6, 7].

A. thaliana is not a metal hyperaccumulator but it is an excellent plant genetic resource to enable more speedy discoveries on this topic [8]. In particular, it is valuable because of the relative ease in isolation of a large number of mutants that can be used for identification and cloning of novel useful genes for metal tolerance, uptake and accumulation or for validation of genes related to these aspects of metal biology. In addition, the novel *Arabidopsis* mutants isolated from studies initially unrelated to heavy metal toxicity and mutant screens involving direct exposure to toxic levels of a heavy metal are excellent genetic tools to evaluate the relative contributions and possible interactions of the different genes related to heavy metal resistance in plants. Selected examples of these studies will be discussed in this chapter to highlight the implications of their findings for phytoremediation strategies.

GENES OF *ARABIDOPSIS THALIANA* RELATED TO METAL RESISTANCE AND UPTAKE

In the plant genome many genes are involved in the detoxification mechanisms for metal homeostasis at the cellular level [9]. For example, several *A. thaliana* genes coding for proteins directly involved in metal transport have been cloned and studied using plant transformation and mutant approaches. Some of the metal detoxification mechanisms may be incompatible with the main purpose of phytoextraction which is to use plants capable of accumulating exceptional levels of soil heavy metals within plant cells without suffering from their toxic effects. It has been revealed that some of these genes may play an important role in regulation of metal resistance but their potential for use in genetic engineering of enhanced phytoextraction of soil heavy metals is doubtful. *AtPRD12* is a good example of this category of metal-related genes. It has been implicated in efflux pumping of Pb at the plasma membrane whereby plant cells can protect the cytoplasm from metal toxicity despite of the presence of increased levels of heavy metals outside [10]. When the *AtPRD12*-knockout plants were grown in media containing 0.5 mM $Pb(NO_3)_2$, the transgenic plants contained more Pb and were less resistant to Pb than WT. In contrast, the *AtPRD12*-overexpressing *A. thaliana* plants accumulated less Pb than WT although they were more resistant to Pb. Therefore, it seems that overexpression of *AtPRD12* in other plants might be of little practical importance as far as phytoextraction of soil Pb is concerned.

Another gene related to metal transport, *AtATM3*, has been shown to be a promising candidate gene for consideration in genetic engineering for enhanced phytoextraction of soil heavy metals. AtATM3 is a mitochondrial protein shown to be implicated in Cd and Pb resistance dependent on the concentration of glutathione [11]. Overexpression of *AtATM3* resulted in enhanced Cd or Pb resistance compared to WT when grown in media containing 40 µM $CdCl_2$ or 0.5 mM $Pb(NO_3)_2$, respectively [11]. More importantly from the perspective of phytoextraction, the increased metal resistance of the *AtATM3*-overexpressing plants was correlated with higher Cd and Pb contents in the shoots of the transgenic plants than WT when grown in the presence of Cd(II) and Pb(II), respectively. This positive feature of overexpression of *AtATM3* was also shown in another plant, *Brassica juncea* (Indian mustard), which is a species already known to be a promising one for phytoremediaton [12]. A new finding emerged from the heterologous overexpression of *AtATM3* was that two metal transporter genes at the plasma membrane and a catalase gene (presumably involved in detoxification of heavy metal-induced oxidative stress) in *B. juncea* were also up-regulated while the expression of another gene (*BjCET2*, cation efflux transporter 2) remained unchanged. Although the precise mechanism linking the heterologous overexpression of *AtATM3* to the differential expression of these *B. juncea* genes is not known, it is possible that overexpression of *AtATM3* could have an indirect effect through regulation of these genes to effectively favor accumulation and tolerance of increased levels of toxic metals inside the cells rather than promoted efflux of the toxic heavy metals. It is also not known if the transfer of *AtATM3* to other non-brassica species would have the same beneficial phytoextraction effect. Nevertheless, it seems a good idea to take into account of this additional possible impact of interaction between transgene expression of *AtATM3* and expression of other endogenous genes in the transgenic plants on the phytoextraction capability of these plants.

Another important metal detoxification strategy is to reduce accumulation of toxic metals in the cytoplasm with the help of the low-molecular weight metal chelating peptides such as phytochelatins (PCs) which are synthesized by the action of PC synthase. These peptides can bind to heavy metals in the cytoplasm forming the PC-metal complexes which are then transported to the vacuole for sequestration with the aid of two vacuole membrane-localized metal transporters, AtABCC1 and AtABCC2 [13]. This PC-dependent metal detoxification pathway has been implicated in increased accumulation and tolerance of Cd and Pb but not Cu in *A. thaliana* [14]. However, increased production of PCs over the level in WT by overexpression of *AtPCS1* (a gene coding for PC synthase) without simultaneous expression of *GSH1* (a gene responsible for the synthesis of glutathione) in transgenic *A. thaliana* resulted in hypersensitivity to Cd stress [14, 15]. Interestingly, the heterologous expression of a PC synthase gene alone from the sacred lotus (*Nelumbo nucifera*) or the simultaneous expression of a PCS gene from garlic and a GSH1 gene from baker's yeast in transgenic *A. thaliana* plants resulted in elevated production of PCs, increased Cd tolerance and accumulation [15, 16]. Therefore, stacking the genes (from different sources) involved in the PC-dependent metal detoxification pathway may also be a useful strategy to enhance phytoextraction of soil metals.

There is some evidence that the vacuole membrane-localized metal transporters, AtABCC1 and AtABCC2, are key players in metal tolerance and accumulation at least in *A. thaliana* [13]. Double knock-out mutants of these two genes resulted hypersensitivity to Cd and accumulation of Cd in the cytoplasm rather than in the vacuole as found in WT [13]. Overexpression of *AtABCC1* alone has been linked to increased tolerance to and accumulation of Cd in the vacuole of the transgenic *A. thaliana* plants. Besides, *AtABCC1*-overexpressing *T. thaliana* plants are also tolerant to arsenic (As) and mercury (Hg). It is also of interest that a putative function of AtABCC1 and AtABCC2 seems to be also involved in the regulation of the transport of Cd between shoot and root since in the double knockout mutants of these two genes there was more Cd transport from the root to the shoot compared to the wild-type. Further work on the mechanism involved might shed more lights on a possible genetic manipulation strategy to enhance transport of heavy metals to the shoots, a desirable trait for genetic engineering to improve phytoextraction of soil heavy metals.

GENES INITIALLY CHARACTERIZED IN NON METAL-RELATED STUDIES

Heavy metal stress is another form of abiotic stress and plants might share some common response pathway to different forms of abiotic stress [17]. It seems likely to find a gene related to regulation of metal resistance in plants from looking at genes and mutants isolated in studies on other forms of abiotic stress such as low temperature, salinity *etc*. [18, 19]. For example, it was first shown that expression of the *ADS2* (*Arabidopsis desaturase 2*) gene was promoted by a low temperature stress treatment [19]. In a later study, when 2-week-old *A. thaliana* seedlings were treated with 50 µM $CdCl_2$ or 0.5 mM $Pb(NO_3)_2$, the transcript level of the gene was greatly reduced in the Cd(II) but not Pb(II) treatment compared to control, suggesting for the first time that expression of the *ADS2* gene is linked to the regulation of Cd resistance in plants. The gene is probably not useful for phytoextraction of Cd or Pb as a T-DNA insertion mutant (*ads2*) with no detectable level of the *ADS2* transcript exhibited enhanced resistance to 50 or 75 µM $CdCl_2$ but a lower level of Cd in both the root and shoot. Interestingly, there was an accompanying increase in the transcript levels of *AtPDR8, GSH1* and *AtATM3*. This suggests that efflux pump activity was favored more in the *ads2* mutant over the potential contribution for vacuolar sequestration *via* the GSH-mediated pathway or accumulation of Cd by the up-regulation of *AtATM3*.

EIN2

The *Arabidopsis Ethylene-Insensitive* 2 gene originally identified in a loss-of-function mutant (*ein2*) that is strongly insensitive to exogenous ethylene [20]. EIN2 is an important component of the ethylene response transduction pathway. It has since been studied in relation to hormonal interrelationships as well as implicated to be involved in many other forms of abiotic and biotic stress including high salinity, oxidative and disease resistance [21]. Its connection to heavy metal resistance was extended in a recent study showing that the transcript level of this gene was greatly elevated in *A. thaliana* seedlings exposed to 0.5 mM $Pb(NO_3)_2$ [20]. The *ein2* mutant plants were more sensitive to 0.5 mM $Pb(NO_3)_2$ and accumulated more Pb than WT. The loss of sensitivity to exogenous ethylene in this mutant was also associated with a reduction in the transcript level of a metal efflux gene (*AtPDR12*) and GSH synthesis linked to the metal detoxification ability (vacuole sequestration) [20]. The molecular mechanisms linking the ethylene response pathway to the expression of genes related to metal resistance is not known. It remains to be seen if other metal uptake and detoxification genes are also affected or not in the *ein2* mutant. The possibility that there is a cross talk between regulation of metal-related genes and other plant hormones (for example, brassinosteroids in Cd response in *A. thaliana*, [22]) involved in many forms of abiotic stress needs to be investigated to a greater depth from the perspective of genetic engineering for enhanced phytoextraction of soil heavy metals.

ACBP1

Acyl-CoA-binding proteins (ACBPs) are implicated in acyl-CoA transport and lipid metabolism and in *A. thaliana* ACBP1 is a membrane-associated protein [23]. A new property of ACBP1 was revealed in an *in vitro* gel assay showing that it was capable of binding Pb(II) [23]. The possibility that it was involved in Pb tolerance and accumulation in *A. thaliana* was then suggested and confirmed. Overexpression of *ACBP1* in transgenic *A. thaliana* plants made them more tolerant to Pb(II)-induced growth inhibition than WT. Moreover, the transgenic plants accumulated more Pb in the shoots than WT when the roots were exposed to 1 mM $Pb(NO_3)_2$ for 48 h. There was no significant difference in the Pb content of the roots between the transgenic plants and WT. It has been postulated that

ACBP1 is probably involved in the uptake of Pb through the plasma membranes into *A. thaliana* tissues and translocation from the root to the shoot. Although the precise mechanism how ACBP1 carries out this function remains to be elucidated, *ACBP1* is a candidate gene for transfer to other plants for manipulation of their Pb phytoextraction capability.

MUTANTS ISOLATED FOLLOWING DIRECT EXPOSURE TO ELEVATED LEVELS OF A HEAVY METAL

In a primary screen intended to isolate novel Cd-resistant mutants, EMS-mutagenized M_2 *Arabidopsis thaliana* (Col-0) seeds were germinated on a culture medium supplemented with 100 μM $CdCl_2$ for 10 days. One of the nine seedlings out of approximately 10,000 seedlings screened survived this treatment [24]. From this initial screen, a line of novel Cd-resistant mutant *A. thaliana* plants was subsequently obtained and named as *cdr3-1D* which exhibited increased root length and fresh weight gain when grown for 3 weeks in 75 μM $CdCl_2$ or 0.5 and 0.75 mM $Pb(NO_3)_2$, or 50 and 75 μM $CuSO_4$ compared to WT. In a related screening program, about 8,600 EMS-mutagenized M_2 *A. thaliana* (Col-0) seeds were germinated on a culture medium supplemented with 50 μM $CdCl_2$ for 10 days. Then seven Cd-sensitive seedlings with small and yellow leaves were selected for further investigation leading to the isolation of a mutant named as *cms1-1* which was more sensitive to 75 μM $CdCl_2$ but was resistant to 0.5 mM $Pb(NO_3)_2$ than WT [25]. Use of this sort of mutants isolated in screens involving the direct exposure of seedlings to toxic levels of heavy metals should be very valuable to assess the relative significance of the different genes shown to be involved in metal resistance in previous studies. In particular, the significance of the genes for phytoremediaton purposes can be better validated in this way than when often a particular gene was studied singly regarding their involvement in metal resistance.

Enhanced Cd-resistance of *cdr3-1D* plants was shown to be caused by a single dominant nuclear gene mutation whereas increased Cd-sensitivity of *cms1-1* was linked to a single recessive nuclear gene mutation. Superficially, the mutated gene in *cdr3-1D*, if cloned, would be expected to be of interest for genetic engineering of plants for phytoremediation as the transfer of a single gene, not multiple genes might achieve improved Cd, Pb and Cu resistance in plants. Since multiple heavy metal contaminants including Cd and Pb are often co-present in polluted soils, this mutated gene would be one of the genes of choice and very useful for plants to be deployed for phytoremediation of heavy metal-polluted soils. However, upon closer scrutiny it appears that *cdr3-1D* plants accumulated less Cd and Pb than WT when grown in 75 µM $CdCl_2$ and 0.75 mM $Pb(NO_3)_2$, respectively. Therefore, transfer or overexpression of this mutated gene into other plants might lead to Cd and Pb resistance based on reduced Cd and Pb uptake. This would be contrary to the purpose of phytoremediation, particularly increased accumulation or phytoextraction of heavy metal contaminants from the polluted environment.

In *cdr3-1D* the mutated gene appeared to increase the expression of several genes known to be associated with Cd resistance including *AtPDR8* and *AtPDR12* (efflux pumps to exclude Cd). This is consistent with the reduced Cd content when the mutant plants were grown in the presence of 75 µM $CdCl_2$ [24]. Interestingly, the expression of *AtATM3* was also elevated which would act in the opposite direction of the efflux pumps and tend to enhance accumulation of Cd and Pb in the shoot and hence increase tolerance [11]. Thus the *cdr3-1D* mutation would seem to act upstream in the regulation of these genes [24]. In addition, these gene expression patterns suggest that concomitant up-regulation of efflux pumps could override the possible contribution of *AtATM3* so that the phenotype of *cdr3-1D* seems to just reflect the action of the efflux pumps only.

CONCLUSIONS

A. thaliana is a valuable genetic resource to draw on for the development of enhanced phytoextraction of soil heavy metals. The *A. thaliana* mutants whether isolated in screens involving direct exposure to heavy metals or *via* other means such as random T-DNA insertion should be used to comprehensively monitor all the genes known to be involved in metal uptake, efflux, detoxification including vacuole sequestration, transport from root to shoot. The interaction of a metal-related transgene and possible impact of the transgene on the expression of all the known genes involved in metal uptake, tolerance and transport in the transgenic plants would also need to be studied. In addition, the impact on endogenous hormonal signal transduction pathways, particularly those known to be involved in abiotic stress has to be evaluated as metal toxicity stress is likely to be interwined with these pathways.

ACKNOWLEDGEMENTS

Declared none.

CONFLICT OF INTEREST

The author(s) confirm that this chapter content has no conflict of interest.

REFERENCES

[1]   Koornneef M, Meinke D. The development of *Arabidopsis* as a model plant. Plant J 2010; 61: 909-21.

[2]   De Pessemier J, Chardon F, Juraniec M, Delaplace P, Hermans C. Natural variation of the root morphological response to nitrate supply in *Arabidopsis thaliana*. Mehanisms Develop 2013; 130: 45-53.

[3]   Chaffai R, Koyama H. Heavy metal tolerance in *Arabidopsis thaliana*. In: Kalder JC, Delseny M, Eds. Adv Bot Res, 2011; 60: 1-49.

[4]   Rascio N, Navari-Izzo F. Heavy metal hyperaccumulating plants: How and why they do it? And what makes them so interesting? Plant Sci 2011; 180: 169-81.

[5]   Bhargava A, Carmona FF, Bhargava M, Srivastava S. Approaches for enhanced phytoextraction of heavy metals. J Environ Manag 2012; 105: 103-20.

[6]   Kraemer U. Metal hyperaccumulation in plants. Ann Rev Plant Biol 2010; 61: 517-34.

[7]   Seth CS. A review on mechanisms of plant tolerance and role of transgenic plants in environmental clean-up. Bot Rev 2012; 78: 32-62.

[8]   Phang IC, Leung DWM, Taylor HH, Burritt DJ. The protective effect of sodium nitroprusside (SNP) treatment on *Arabidopsis thaliana* seedlings exposed to toxic level of Pb is not linked to avoidance of Pb uptake. Ecotoxicol Environ Safety 2001; 74: 1310-5.

[9]   Kobayashi T, Nishizawa NK. Iron uptake, translocation, regulation in higher plants. Ann Rev Biol 2012; 63: 131-52.

[10]   Lee M, Lee K, Lee J, Noh EW, Lee Y. AtPDR12 contributes to lead resistance in *Arabidopsis*. Plant Physiol 2005; 138: 827-36.

[11]   Kim DY, Bovet L, Kushnir S, Noh EW, Martinoia E, Lee Y. AtATM3 is involved in heavy metal resistance in *Arabidopsis*. Plant Physiol 2006; 140: 922-32.

[12]   Bhuiyan MSU, Min SR, Jeong WJ, *et al.* Overexpression of *AtATM3* in *Brassica juncea* confers enhanced heavy metal tolerance and accumulation. Plant Cell Tiss Org Cult 2011; 107: 69-77.

[13]   Park J, Song WY, Ko D, *et al.* The phytochelatin transporters AtABCC1 and AtABCC2 mediate tolerance to cadmium and mercury. Plant J 2012; 69: 278-88.

[14] Lee S, Petros D, Moon JS, Ko TS, Goldsbrough PB, Korban SS. Higher levels of ecotypic expression of *Arabidopsis* phytochelatin synthase do not lead to increased cadmium tolerance and accumulation. Plant Phyiol Biochem 2003; 41: 903-10.

[15] Guo J, Dai X, Xu W, Ma M. Overexpressing GSH1 and AsPCS1 simultaneously increases the tolerance and accumulation of cadmium and arsenic in *Arabidopsis thaliana*. Chemosphere 2008; 72: 1020-6.

[16] Liu ZL, GU CS, Chen FD, *et al.* Heterologous expression of a *Nelumbo nucifera* phytochelatin synthase gene enhances cadmium tolerance in *Arabidopsis thaliana*. App Biochem Biotech 2012; 166: 722-34.

[17] Winter TR, Borkowski L, Zeier J, Rostas M. Heavy metal stress can prime for herbivore-induced plant volatile emission. Plant Cell Environ 2012; 35: 1287-98.

[18] Cao S, Bian X, Jiang S, Chen Z, Jian H, Sun Z. Cold treatment enhances Pb resistance in *Arabidopsis*. Acta Physiol Plant 2010; 32: 19-25.

[19] Yu B, Bian X, Qian JZ, Chen XP, Wang RF, Cao SQ. *Arabidopsis* desaturase 2 gene is involved in the regulation of cadmium and lead resistance. Plant Soil 2012; 358: 289-300.

[20] Cao SQ, Cao S, Chen ZY, *et al.* The *Arabidopsis* ethylene-insensitive 2 gene is required for lead resistance. Plant Physiol Biochem 2009; 47: 308-12.

[21] Wang YN, Liu C, Li KX, *et al.* *Arabidopsis* EIN2 modulates stress response through abscisic acid response pathway. Plant Mol Biol 2007; 64: 633-44.

[22] Villiers F, Jourdain A, Bastien A, *et al.* Evidence for functional interaction between brassinosteroids and cadmium response in *Arabidopsis thaliana*. J Exp Bot 2012; 63: 1185-200.

[23] Xiao S, Gao W, Chen QF, Malingam S, Chye ML. Overexpression of membrane-associated acyl-CoA-binding protein ACBP1 enhances lead tolerance in *Arabidopsis*. Plant J 2008; 54: 141-51.

[24] Wang Y, Zhong K, Jiang L, Sun, Y. Characterization of an *Arabidopsis* cadmium-resistant mutant *cdr3-1D* reveals a link between heavy metal resistance as well as seed development and flowering. Planta 2011; 233: 697-706.

[25] Lv SC, Sun ZH, Qian JZ, *et al.* Isolation and characterization of a novel cadmium-sensitive mutant in *Arabidopsis*. Acta Physiol Plant 2012; 34: 1107-18.

CHAPTER 7

Phytomanagement of Contaminated Sites Using Poplars and Willows

Brett Robinson[1,*] and Ian McIvor[2]

[1]Department of Soil Science, Lincoln University, P.O. Box 7647, Lincoln, New Zealand and [2]Plant and Food Research, Private Bag 11030, Palmerston North, New Zealand

Abstract: Poplars and willows find widespread use for the phytomanagement of contaminated sites. We review the processes involved in poplar and willow phytomanagement with the aim of elucidating knowledge gaps and fertile areas for future research. Poplars and willows are fast-growing tree species reduce contaminant mobility by their high water use and stabilisation of the contaminated substrate. Root exudates and microorganisms associated with the rhizosphere of these trees promote the degradation of some organic contaminants. Some clones accumulate of toxic trace elements, such as Cd, thereby facilitating their entry into the food chain. Phytomanagement using poplars and willows can result in biomass products, such as timber, bioenergy, essential oils, organic mulches, and stock fodder. These offset the cost of the operation. There is great potential to improve the performance of phytomanagement systems using these species by exploiting their natural genetic variation and microbiological symbionts. A key unknown is the response of deep-rooted poplars to heterogeneities in soil, which may promote or reduce contaminant mobility.

Keywords: Avoidance, biomass products, clonal variation, dairy effluent, deep-rooted species, endophytic bacteria, evapotranspiration, food chain, genetic variation, groundwater, landfill, leaching, merA or merB genes, phytomanagement, phytostablilization, poplars, *Salix purpurea*, *Salix viminilis*, trace elements, willows.

INTRODUCTION

Phytomanagement describes the engineering or manipulation of soil-plant systems to control the fluxes of contaminants in the environment [1]. Successful

---

*Address correspondence to Brett Robinson: Department of Soil Science, Lincoln University, P.O. Box 7647, Lincoln, New Zealand; Tel: +64 21 288 5655; Fax: +64-3-325-3607; E-mail: brett.robinson@lincoln.ac.nz

phytomanagement should either cost less than other remediation, or be a profitable operation, by producing valuable plant biomass products. This may include bioenergy or timber production on contaminated land, a practice that does not reduce food production.

Poplars and willows are fast-growing phreatophytic tree species with a worldwide distribution that are of widespread use for phytomanagement. In this role, the trees function as bio-pumps that use the sun's energy to dewater contaminated sites and control leaching, as well as enhance the organic matter and microbial activity in the rhizosphere [2]. These root-zone processes augment contaminant degradation and reduce leaching. Poplars and willows are therefore suited to treating a wide range of contaminants in numerous media.

Poplars and willows have far greater potential for stabilisation of contaminated sites than the now debunked concept of heavy- metal phytoextraction from soils. Poplars are particularly suited to remediation of organochlorines, polycyclic aromatic hydrocarbons (PAHs), excess ammonia and nitrates, and the immobilisation of heavy-metal contaminants. Here, we review the key processes in contaminant phytomanagement using poplars and willows with the aim of identifying key knowledge gaps and fertile areas for future research.

THE ROLE OF POPLARS AND WILLOWS IN PHYTOMANAGEMENT

Phytomanagement of contaminated sites requires that the site be vegetated as quickly as possible at minimum cost. Whereas many tree species require careful propagation in nurseries and transplantation at the contaminated site, the propagation of poplars and willows usually involves the insertion of cuttings into the ground [3]. This is a rapid and low-cost procedure [4]. Cuttings can be produced using stool beds in nurseries, with no need to raise individual seedlings. The growth and survivability of the cuttings is proportional to their size, which typically ranges between 0.3 m - 3 m in length [5].

Asexually propagated poplar and willow cuttings are genetically identical. This allows the selection of clones with the best characteristics to phytomanage any particular site [5]. Various clones can tolerate a wide range of environmental conditions, especially with regard to water availability [6]. There is a wide variation in the ability of different willow clones to accumulate heavy metals in the bark and wood [7, 8]. There is inter-clonal variation in tolerance to Cu, Cd, Ni and Zn, as well as in the uptake of Cu [5]. Large genetic differences were found in the uptake of trace elements by willow clones [9]. In the case of Cd, there was a 12-fold difference between the lowest and highest leaf Cd concentrations for sibling plants growing in identical environments. Ideally, several clones or a species mixture should be employed, lest a change in environmental conditions, such as the appearance of a plant pathogen results in plant dieback across the entire site. An example of such an event was the appearance in New Zealand of the invasive pest the willow saw-fly (*Nematus oligospilus*) in the late 1990s [10]. This completely defoliated some willows clones causing flooding in areas where sensitive clones were used exclusively for riverbank stabilisation.

The rapid growth rate of poplars and willows reduces the need for weed control at the site, with dense plantings producing a closed-canopy cover within three years [11]. This trait promotes rapid greening of the site. Hydraulic control and surface stabilisation occur more rapidly with poplars and willows compared to most other tree species, which grow more slowly. When planted along riverbanks, poplars and willows can provide effective barrier contaminants in runoff or subsurface flow. They also provide a physical barrier to prevent stock from entering the waterways, where they can cause damage from trampling and contamination by urination and defecation. The abundance and diversity of benthic invertebrates have been shown to be grater in willow-protected streams [12]. The authors also reported greater numbers of large brown trout in willowed streams. However, in small streams, willows can reduce biodiversity due to shading out of aquatic flora [13].

During periods of drought, deep-rooted poplars and willows have greater access to water and continue to transpire after drought renders shallow-rooted species dormant. It was shown that evaporation for widely spaced poplar-pasture systems was significantly greater than pasture alone [14]. The canopies of poplar and willow stands act as umbrellas. Depending on climatic conditions, more than 15% of rainfall may evaporate before it reaches the ground [15]. On disused tips and landfill sites, plantings of poplars and willows can prevent contaminants leaching to receiving waters [16]. In addition to providing hydraulic control on landfills, poplars and particularly willows may be used to treat landfill leachates, although there is a wide range in efficacy between contaminants [17]. The installation of a tree well system [18] can force poplar roots down to a depth of several metres, where they can interact with groundwater. Thus, the trees may not only be used to manage groundwater flows, but also interact with contaminants in the groundwater itself.

Poplars and willows have a high fine root density [19], which leads to increased contact with soil contaminants, decreasing the negative impacts of contaminant heterogeneity on remediation time. Root exudates enhance biological activity, which in turn, can result in the accelerated degradation of some organic contaminants [20]. Biological activity in the rootzone of willows is significantly greater than perennial ryegrass [21]. Their roots penetrate further than grasses and improve soil aeration due to high water use, further promoting microbiological activity. A range of aromatic and aliphatic petroleum hydrocarbons were found to be effectively degraded (<25% remaining in topsoil) under a poplar tree phytoremediation system in a disused oil tank farm [22]. Poplars could reduce the concentrations of chloroethylenes in soil water by 99% [23]. Degradation was found to be complete, with 95% of the recovered chlorine present as chloride. A key unknown here is how poplar roots react to heterogeneity in soil. Root avoidance of contaminant hotspots will reduce any potential positive effect that the trees may have on contaminant degradation. Conversely, in soils where the roots promote preferential flow pathways [24], root avoidance may result in lower water flux through contaminant rich zones and hence reduce contaminant leaching.

One of the drawbacks of using poplars and willows in sites contaminated with volatile organic compounds such as TCE (trichloroethylene) is that the trees may volatilise these toxic compounds into the atmosphere, thereby contaminating adjacent areas. A possible solution to this problem is the inoculation of genetically modified endophytic bacteria, such as *Pseudomonas putida,* that metabolise volatile contaminants into more benign compounds [25]. Research into the suite of endophytic bacteria in poplar may improve the not only the growth of these trees on contaminated sites, but also the way in which contaminants are metabolised or accumulated [26].

Compared to other species, poplars and willows translocate high concentrations of trace elements to their above-ground portions. This has been attributed to their low water-use efficiency [21]. Accumulation of contaminating trace elements such as B may reduce their downward mobility, thereby protecting groundwater [11]. The ability of willows to accumulate Cd has been suggested as a means to exploit Cd from contaminated soils [27-29]. However, to date, there are no commercial full-scale Cd phytoextraction operations.

The accumulation of contaminating trace elements may facilitate their entry into the food chain. Close monitoring of Cd concentrations in poplars was required in the Guadiamar phytomanagement programme due to concerns that this element may enter the food chain *via* leaf herbivory [30]. Moreover, since these elements may be accumulated at depth, the trees, through leaf abscission, may effect a redistribution of these elements to the surface horizons [31], where they are more likely to come into contact with other soil fauna and flora.

Poplars and willows can aid the establishment of a natural ecosystem on a contaminated site, where the value of phytomanagement comes from increased biodiversity, improved aesthetics and as a recreational area. Here the poplars and willows may be valuable in their own right or as a primary vegetation cover into which other native species can be planted. However, in some countries such as Australia and New Zealand, the public views of poplars and, in particular, willows as invasive alien weeds [32], detract their use in phytomanagement.

BIOMASS PRODUCTS FROM POPLARS AND WILLOWS DURING PHYTOMANAGEMENT

Poplar and willow material can be used for bioenergy, pulp & paper, solid & composite wood products, and feed products [33] as well as a nutrient-rich organic mulch [11]. This is important for price sensitive markets because secondary products offset the costs of remediation. Harvested material could be converted to energy in an incinerator equipped with a Cottrell precipitator [34] that prevents any accumulated contaminants being lost in the smoke. The energy produced by the incineration of the crop might even be utilised for generating electricity [35]. Flash pyrolysis of willow biomass could produce oils. At relatively low temperatures, most heavy metal contaminants enter the ash phase, with negligible amounts being volatilised or entering the oil [36].

Provided the contaminant concentrations in the above-ground portions are not excessive, poplars and willows can provide supplementary stock fodder [32]. Both foliage and small twigs can be browsed by sheep (Fig. **1**) and cattle [37, 38]. Poplars and willows as stock feed have proven health benefits such as increased fecundity [39]. This has been attributed to the high protein [40] and tannin content of the leaves [41].

Figure 1: The leaves and twigs of willows and (in this case) poplars make palatable stock fodder.

Poplars and willows accumulate high concentrations of Zn and Co relative to pasture species and this may be linked to observed health benefits in sheep and cattle when foliage is used as stock fodder [42]. A potential problem of using poplars and willows as stock fodder at some sites is their accumulation of Cd [9], although varieties exist that accumulates high Zn and Co, but low Cd [42].

Using poplars and willows for biomass products has the advantage that the above-ground portions can be excised without killing the tree. There is no need to replant the site because new branches sprout from the stump or trunk [7]. Coppicing (cutting from the base of the trunk) or pollarding (cutting only the portions >1- 2 m above ground) eliminates evapotranspiration and therefore reduces the efficacy of the trees to phytomanage a contaminated site, particularly if hydraulic control is a primary objective. A solution is partial coppicing, where selected rows of trees are not harvested, thus retaining some hydraulic control on the site. For example, in a densely planted stand, coppicing alternate rows of trees (*i.e.,* 50% of the total biomass on a site) can reduce evapotranspiration by as little as 20% [11] because the remaining trees intercept more light.

CASE STUDIES

Phytomanagement of Dairy Effluent

Dairy farming in New Zealand requires the disposal of large volumes of N-rich effluent, which is often pumped onto pasture to improve growth. This can result in N contamination of receiving waters. Small blocks of poplars and willows can be more effective than pasture at reducing N-leaching because of their deeper root systems and re-evaporation of some incident rainfall [43]. Poplar biomass can be increased threefold upon the addition of dairy effluent (Fig. **2**). This rapid growth results lignified tissue that is soft, with a low-density. While this is unsuitable for wood production, the soft tissues may improve palatability for stock fodder. Coppicing can occur once or twice a year, and the biomass used as a feed supplement or an N-rich organic mulch.

Phytomanagement of a Contaminated Wood-Waste Pile

Many New Zealand timber products are treated with pentachlorophenol and chromated copper arsenate (CCA) to prevent decay. Boron is used to prevent sap stain. Treatment sites and wood-waste disposal sites have become contaminated and pose a risk to receiving waters through leaching. The Kopu timber-waste at the base of the Coromandel peninsula, New Zealand (37.2° S, 175.6° E) has a surface area of 3.6 ha and an average depth of 15 m. Sawdust and yard-scrapings were dumped for some 30 years from 1966. Geotechnical engineering ensures no surface or ground water enters the pile. Leachate resulting from rainfall is collected in a holding pond. Vegetation has failed to establish naturally. Before the implementation of phytomanagement, the evaporation from pile was negligible. The annual rainfall 1135 mm caused regular leaching into a local stream. Boron-rich leachate raised the stream B concentration to >1.4 mg L^{-1}, the New Zealand Drinking Water Standard (NZDWS), especially in the summer months when stream flow was low.

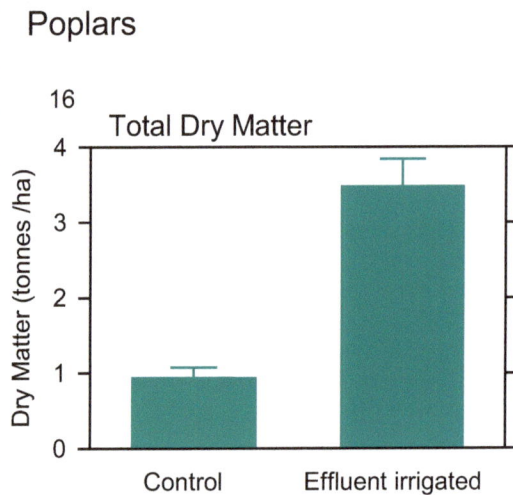

Figure 2: Effect of dairy shed effluent on poplar biomass production.

In July 2000, a 1 ha trial was set up using 10 poplar and willow clones and two species of *Eucalyptus*. Two *Populus deltoides* hybrid clones were best candidates for phytoremediation based on survival, biomass production and B uptake. There were large variations in the growth of poplars and willows in the wood-waste (Fig. **3**). In July 2001, the remainder of the site was planted at a density of 7000 trees ha^{-1}. Fertilisers were added periodically. A pump was installed to recirculate leachate as irrigation during the summer months. Fig. **4** shows tree growth on the Kopu pile after three years.

Before planting the bare pile leached during all months of the year. The trees reduced the drainage to the three winter months. Summer is the greatest concern for the contamination of waterways because low stream flows result in less contaminant dilution. Drainage collected during the winter months is irrigated onto the trees to relieve summer drought. During winter, drainage may be released into a nearby stream at times of high flow when the risk of exceeding the New Zealand Drinking Water Standard (NZDWS) is minimal.

Figure 3: A field trial of poplar and willow clones on a contaminated wood-waste pile after 2 years. Note the range in biomass production.

Figure 4: The Kopu wood-waste pile three-years after planting.

Poplar leaves had average Cu and Cr concentrations of 6.6 and 4.9 mg kg^{-1} dry mass. Arsenic concentrations were below detection limits (1 mg kg^{-1}). These low leaf concentrations will not facilitate the entry of CCA into the food chain. Before abscission, the average leaf-B concentration was nearly 700 mg kg^{-1} on a dry matter basis, over 28 times higher than the B concentration in the sawdust (40 mg kg^{-1} dry matter). Selective coppicing of the trees may allow the production of a B-rich organic mulch that could be applied to nearby orchards that are deficient in B.

Phytoremediation of a Site Contaminated with Organo-Chlorine Pesticides

New Zealand has an estimated 50,000 disused sheep dip sites contaminated with As, organo-chlorines and organophosphates. Land use changes can result in these sites being incorporated into residential or intensive agricultural zones. Groundwater contamination has also been demonstrated [44]. Sites were often installed near wells or streams to facilitate preparation and disposal of the pesticide solution. A disused sheep-dipping site in an asparagus field was discovered following the measurement of elevated dieldrin concentrations in a nearby well. Soil analyses revealed dieldrin concentrations from 10 - 70 mg kg^{-1} over 100 m^2. The Dutch Intervention Value for dieldrin in soil is 4 mg kg^{-1}. In September 2001, the site was planted in willow clones. In Jan 2005, the average height of the trees was over 5 m (Fig. **5**).

Figure 5: Four-year-old willows remediating disused sheep dip site in an asparagus field.

Parallel experiments using soil collected from this dipping site showed that willows effected a 20% degradation over a period of five months. The biological activity, as measured by dehydrogenase activity, was six times higher in the root-zone of willows compared to a pasture species. The effect of the willows on this sheep dip site is to prevent the cultivation and harvesting of asparagus, promote the degradation of dieldrin and reduce dieldrin leaching by enhanced evapotranspiration.

DEVELOPMENT OF POPLAR AND WILLOW PHYTOMANAGEMENT

The considerable genetic variation between poplar and willow varieties allows the creation of new varieties that are optimised for phytomanagement, by selective breeding or genetic engineering. *Salix viminilis* and *S. purpurea* are consistently reported as high accumulators for some metals, especially Zn and Cd [45] and may find use to extract these metals from contaminated sites. Genetic modification to include merA or merB genes induces these plants to accumulate Hg, where the Hg is either volatilised or stored in the leaves [46]. There are several problems with such an approach. There is no control over the final destination of the volatilised Hg, which will simply be incorporated into the soils, waters and biota of surrounding areas. Mercury that is accumulated in the leaves may be consumed by herbivores, thus entering the food chain, where concentrations of this biophillic element may reach toxic concentrations in predators, possibly affecting humans. Leaf fall will result in the accumulation of Hg in the surface layers. For most Hg-contaminated sites, such a phytoextraction operation will not be economically viable for reasons outlined in [1].

Genetic manipulation, or inoculation of the trees with endophytes or rhizobacteria such as *Pseudomonas putida,* may improve the tolerance of poplars to soil contaminants and may aid the degradation of organic contaminants in soils or groundwater [47].

CONCLUSIONS

Although poplars and willows find widespread use in the phytomanagement of contaminated sites, there are several knowledge gaps, which if filled, could greatly enhance the performance of phytomanagement systems. Identifying mechanisms by which these trees promote the degradation of organic contaminants could result in the development of clones that are more efficient in this role. A particularly promising development in this area is the elucidation of the role of bacteria and fungi, which thrive in the root-zone of these trees, on contaminant degradation. Hitherto, it has been assumed that establishing deep-rooted poplar trees on contaminated sites will result in a reduction of contaminant leaching due to increased evapotranspiration. However, macropores eventuating from root development may result in the creation of preferential flow pathways through which particulate-bound contaminants, such as Pb, may travel to groundwater. Similarly, further research on combining poplars and willows with indigenous plant species may create phytomanagement systems that have increased resilience to environmental changes, increased biodiversity and are more appealing to stakeholders.

ACKNOWLEDGEMENTS

Declared none.

CONFLICT OF INTEREST

The author(s) confirm that this chapter content has no conflict of interest.

REFERENCES

[1]   Robinson B H, Banuelos G, Conesa H M, Evangelou M W H, Schulin R. The Phytomanagement of Trace Elements in Soil. Crit Rev Plant Sci 2009; 28: 240-66.

[2]   Robinson B, Green S, Mills T, *et al*. Phytoremediation: using plants as biopumps to improve degraded environments. Aust J Soil Res 2003; 41: 599-611.

[3]   Weyens N, Truyens S, Dupae J, *et al*. Potential of the TCE-degrading endophyte Pseudomonas putida W619-TCE to improve plant growth and reduce TCE phytotoxicity and evapotranspiration in poplar cuttings. Environ Pollut 2010; 158: 2915-9.

[4]   Quinn J J, Negri C M, Hinchman R R, Moos L P, Wozniak J B, Gatliff E. Predicting the effect of deep-rooted hybrid poplars on the groundwater flow system at a large scale phytoremediation site. Internat J Phytoremed 2001; 3: 41-60.

[5]   Punshon T, Dickinson N. Heavy metal resistance and accumulation characteristics in willows. Internat J Phytoremediat 1999; 1: 361-85.

[6]   Fillion M, Brisson J, Teodorescu T I, Sauve S, Labrecque M. Performance of *Salix viminalis* and *Populus nigra* x *Populus maximowiczii* in short rotation intensive culture under high irrigation. Biomass Bioenergy 2009; 33: 1271-7.

[7]   Laureysens I, Blust R, De Temmerman L, Lemmens C, Ceulemans R. Clonal variation in heavy metal accumulation and biomass production in a poplar coppice culture: I. Seasonal variation in leaf, wood and bark concentrations. Environ Pollut 2004; 131: 485-94.

[8]   Riddell-Black D, Pulford I D, Stewart C. Clonal variation in heavy metal uptake by willow. In: Biomass and energy crops. Meeting of the Association of Applied Biologists, 7-8 April 1997, Royal Agricultural College, Cirencester, UK; pp. 327-34.

[9]   Granel T, Robinson B, Mills T, Clothier B, Green S; Fung L. Cadmium accumulation by willow clones used for soil conservation, stock fodder, and phytoremediation. Aust J Soil Res 2002; 40: 1331-7.

[10]   Charles J G, Allan D J. Development of the willow sawfly, Nematus oligospilus, at different temperatures, and an estimation of voltinism throughout New Zealand. N.Z J Zool 2000; 27: 197-200.

[11]   Robinson B H, Green S R, Chancerel B, Mills T M, Clothier B E. Poplar for the phytomanagement of boron contaminated sites. Environ Pollut 2007; 150: 225-33.

[12]   Glova G J, Sagar P M. Comparison of fish and macroinvertebrate standing stocks in relation toriparian willows (salix spp) in 3 new-zealand streams. NZ J Mar Freshw Res 1994; 28: 255-66.

[13]   Lester P J, Mitchell S F, Scott D. Effects of riparian willow trees (salix-fragilis) on macroinvertebrate densities in 2 small central otago, new-zealand, streams. NZ J Mar Fresh Warter Res 1994; 28: 267-76.

[14]   Guevara-Escobar A, Kemp P D, Mackay A D, Hodgson J. Soil properties of a widely spaced, planted poplar (Populus deltoides)-pasture system in a hill environment. Aust J Soil Res 2002; 40: 873-86.

[15]   McNaughton K G, Jarvis P G. Predicting Effects of Vegetation Changes on Transpiration and Evaporation. Water Deficits and Plant Growth. Volume VII. Additional Woody Crop Plants. Minnesota, Academic Press 1983; pp. 1-47.

[16]   Ferro A M, Rieder J P, Kennedy J, Kjelgren R. Phytoremediation of Groundwater Using Poplar Trees. In: Phytoremediation.Thibeault CA, Savage LM, Eds. Southborough, MA, International Business Communications Inc, 1997; pp. 201-12.

[17]   Aronsson P, Dahlin T, Dimitriou I.Treatment of landfill leachate by irrigation of willow coppice - Plant response and treatment efficiency. Environ Pollut 2009; 158: 795-804.

[18]   Roberts E J. Tree well. Available In www.patentgenius.co./patentD562648, USA, 2008.

[19]   Baum C, Leinweber P, Weih M, Lamersdorf N, Dimitriou I. Effects of short rotation coppice with willows and poplar on soil ecology. Landbauforschung Volkenrode 2009; 59: 183-96.

[20]   Susarla S, Medina V F, McCutcheon S C. Phytoremediation: An ecological solution to organic chemical contamination. Ecol Eng 2002; 18: 647-58.

[21]   Mills T, Robinson B, Green S, Clothier B, Fung L, Hurst S. Difference in Cd uptake and distribution within poplar and willow species. In: The 42nd Annual Conference and Expo of the New Zealand Water and Waste Association, Rotorua: New Zealand 2000.

[22]   El-Gendy A S, Svingos S, Brice D, Garretson J H, Schnoor J. Assessments of the efficacy of a long-term application of a phytoremediation system using hybrid poplar trees at former oil tank farm sites. Water Environ Res 2009; 81: 486-98.

[23]   James C A, Xin G, Doty S L, Muiznieks I, Newman L, Strand S E. A mass balance study of the phytoremediation of perchloroethylene-contaminated groundwater. Environ Pollut 2009; 157: 2564-9.

[24] Roulier S, Robinson B, Kuster E, Schulin R. Analysing the preferential transport of lead in a vegetated roadside soil using lysimeter experiments and a dual-porosity model. European J Soil Sci 2008; 59: 61-70.

[25] Weyens N, Van Der Lelie D, *et al.* Bioaugmentation with engineered endophytic bacteria improves contaminant fate in phytoremediation. Environ Sci Techol 2009; 43: 9413-8.

[26] van der Lelie D, Taghavi S, Monchy S, *et al.* Poplar and its Bacterial Endophytes: Coexistence and Harmony. Crit Rev Plant Sci 2009; 28: 346-58.

[27] Greger M, Landberg T. Use of willow in phytoextraction. In: Phytoremediation: How Plants Can Help in Cleaning the Environment. A workshop held during INTECOL 98, the International Conference of Ecology, CRC Press, Florence: Italy 1999; pp. 115-23.

[28] Robinson B H, Mills T M, Petit D, *et al.* Natural and induced cadmium-accumulation in poplar and willow: Implications for phytoremediation. Plant Soil 2000; 227: 301-6.

[29] Witters N, Van Slycken S, Ruttens A, *et al.* Short-Rotation Coppice of Willow for Phytoremediation of a Metal-Contaminated Agricultural Area: A Sustainability Assessment. Bioenergy Res 2009; 2: 144-52.

[30] Dominguez M T, Maranon T, Murillo J M, Schulin R, Robinson B H. Trace element accumulation in woody plants of the Guadiamar Valley, SW Spain: A large-scale phytomanagement case study. Environ Pollut 2008; 152: 50-9.

[31] Vandecasteele B, Quataert P, Genouw G, Lettens S, Tack F M G. Effects of willow stands on heavy metal concentrations and top soil properties of infrastructure spoil landfills and dredged sediment-derived sites. Sci Total Environ 2009; 407: 5289-97.

[32] Wilkinson A G. Poplars and willows for soil erosion control in New Zealand. Biomass Bioenergy 1999; 16: 263-74.

[33] Licht L A, Isebrands J G. Linking phytoremediated pollutant removal to biomass economic opportunities. Biomass Bioenergy 2005; 28: 203-18.

[34] Miloua F, Kadous N, Rahou F Z, Semmah A, Tilmatine A. Granulometric and physicochemical effect of solid particles on filtration efficiency of an electrostatic precipitator of a "Cottrell" type. Mater Technol 2005; 20: 196-9.

[35] Nixon D J, Stephens W, Tyrrel S F, Brierley E D R. The potential for short rotation energy forestry on restored landfill caps. Bioresource Technol 2001; 77: 237-45.

[36] Stals M, Thijssen E, Vangronsveld J, Carleer R, Schreurs S, Yperman J. Flash pyrolysis of heavy metal contaminated biomass from phytoremediation: Influence of temperature, entrained flow and wood/leaves blended pyrolysis on the behaviour of heavy metals. J Analy Appl Pyrolysis 2010; 87: 1-7.

[37] Douglas G B, Bulloch B T, Foote A G.Cutting management of willows (*Salix* spp) and leguminous shrubs for forage during summer. NZ J Agric Res 1996; 39: 175-84.

[38] Hathaway R L. Willows for the Future. National Water and Soil Conservation. Streamland 1987.

[39] Pitta D W, Barry T N, Lopez-Villalobos N, Kemp P D. Willow fodder blocks - An alternate forage to low quality pasture for mating ewes during drought? Anim Feed Sci Technol 2007; 133: 240-58.

[40] Nelson N D, Sturos J A, Fritschel P R, Satter L D. Ruminant feed stuff from commercial foliage of hybrid poplars grown under intensive culture. Forest Products J 1984; 34: 37-44.

[41]   Barry T, Kemp P. Ewes respond to poplar feed. Tree Feed 2001; pp. 2-3.

[42]   Robinson B, Mills T, Green S, Chancerel B, Clothier B, Fung L, Hurst S, McIvor I. Trace element accumulation by poplars and willows used for stock fodder. N Z J Agric Res 2005; 48: 489-97.

[43]   Roygard J K F, Green S R, Clothier B E, Sims R E H, Bolan N S. Short rotation forestry for land treatment of effluent: a lysimeter study. Aust J Soil Res 1999; 37: 983-91.

[44]   Robinson P. The fate of Vetrazin (cyromazine) during wool scouring and its effects on the aquatic environment. In Soil Science. Lincoln University, Lincoln 1995.

[45]   Mleczek M, Kaczmarek Z, Magdziak Z, Golinski P K. Hydroponic estimation of heavy metal accumulation by different genotypes of *Salix*. J Environ Sci Health Part A-Toxic/Hazard Substances & Environ Eng 2010; 45: 569-8.

[46]   Ruiz O N, Daniell H. Genetic engineering to enhance mercury phytoremediation. Curr Opinion Biotechnol 2009; 20: 213-9.

[47]   Xin G, Zhang G Y, Kang J W, Staley J T, Doty S. A diazotrophic, indole-3-acetic acid-producing endophyte from wild cottonwood. Biol Fertility Soils 2009; 45: 669-74.

Index

A

abiotic stress, 56, 62

acyl-CoA-binding proteins (ACBPs), 114

alleviation of metal toxicity, 56, 59

aluminum (Al), 68

apoplastic, 49

Arabidopsis desaturase 2 gene, 113

Arabidopsis Ethylene-Insensitive 2 gene, 114

Arabidopsis halleri, 91

Arabidopsis thaliana, 60, 64, 65, 91, 94, 109

arsenic (As), 68, 70, 72

Athyrium yokoscense, 69

AtABCC1, 112, 113

AtABCC2, 112, 13

AtATM3, 111, 116

ATP-ase, 95

AtPDR8, 113

AtPRD12, 111, 112, 114

avoidance, 122

B

bioaccumulation, 8

biodegradable chelators, 40, 47

bioavailability (of heavy metals), 4, 22, 23, 40, 47

biomass products, 119, 120, 123

bioremediation, 8

biosorption, 8

Brassica juncea, 111

Brassicaceae, 91

C

cadmium (Cd), 68, 92, 93, 94, 110, 121, 123